Collector's Guide to the
BLACK TOURMALINE
of Pierrepont, New York

Steven C. Chamberlain   *Jeffrey R. Chiarenzelli*
George W. Robinson   *Marian V. Lupulescu*
Michael R. Walter   *David G. Bailey*

4880 Lower Valley Road • Atglen, PA 19310

Other Schiffer Books by Steven Chamberlain
Collector's Guide to the Minerals of New York State,
ISBN 978-0-7643-4334-6
Other Schiffer Books on Related Subjects:
Collector's Guide to the Tourmaline Group, Robert Lauf,
ISBN 978-0-7643-3775-8
Collector's Guide to Herkimer Diamonds, Michael R. Walter,
ISBN 978-0-7643-4710-8

Copyright © 2016 by Steven Chamberlain

Library of Congress Control Number: 2016947122

All rights reserved. No part of this work may be reproduced or used in any form or by any means—graphic, electronic, or mechanical, including photocopying or information storage and retrieval systems—without written permission from the publisher.

The scanning, uploading, and distribution of this book or any part thereof via the Internet or any other means without the permission of the publisher is illegal and punishable by law. Please purchase only authorized editions and do not participate in or encourage the electronic piracy of copyrighted materials.

"Schiffer," "Schiffer Publishing, Ltd.," and the pen and inkwell logo are registered trademarks of Schiffer Publishing, Ltd.

Cover design by Molly Shields
Cover Photo: Tourmaline, 5 cm. Collected from the Hillside site by Mike Walter in the summer 1993; now in Jay Walter collection. MW

Type set in BauerBodni BT/Times New Roman

ISBN: 978-0-7643-5199-0
Printed in China

Published by Schiffer Publishing, Ltd.
4880 Lower Valley Road
Atglen, PA 19310
Phone: (610) 593-1777; Fax: (610) 593-2002
E-mail: Info@schifferbooks.com
Web: www.schifferbooks.com

For our complete selection of fine books on this and related subjects, please visit our website at www.schifferbooks.com. You may also write for a free catalog.

Schiffer Publishing's titles are available at special discounts for bulk purchases for sales promotions or premiums. Special editions, including personalized covers, corporate imprints, and excerpts, can be created in large quantities for special needs. For more information, contact the publisher.

We are always looking for people to write books on new and related subjects. If you have an idea for a book, please contact us at proposals@schifferbooks.com.

CONTENTS

4 *Foreword*

5 *Acknowledgments*

6 *Introduction*

8 *Chapter 1. History*
   8   Ownership History
   9   Collector History

25 *Chapter 2. Geology*
   25   Introduction and Tectonic Setting
   26   Geology of the Adirondack Region
   27   Detailed Geology of the Powers Farm Locality
   29   Age of Primary Mineralization
   29   Origin of the Powers Farm Black Tourmaline Locality
   30   Regional Considerations of Powers Farm Study
   30   Summary

32 *Chapter 3. The Sites and Their Minerals*
   33   Top of the Hill
   57   Smoky Quartz Vein
   70   Streamside Veins
   70   Phosphate Vein
   88   Middle Vein
   91   Waddell Vein
   98   Wallace-Carlin Vein
   103 Hillside
   107 Marsh

109 *Chapter 4. Gallery*

114 *Chapter 5. Special Focus Topics*
   114 Tourmaline—Composition and Nomenclature
   117 Tourmaline—Crystal Forms and Habits
   119 Habits of Quartz
   120 Selected Pseudomorphs
   124 Rare-earth Minerals
   125 Periods of Mineralization: Precambrian and Late-Stage

127 *Glossary*

FOREWORD

FROM THE SERIES EDITOR

The black tourmaline occurrence at Pierrepont, New York, is a classic locality in the fullest sense: It has produced a huge quantity of fine specimens represented in museum collections internationally and in the collections of mineral enthusiasts of more modest means. It has been exploited for almost 150 years, is still producing today, and is accessible to collectors as a fee site. Moreover, the locality is scientifically fascinating as geologists continue to study the complex suite of associated minerals and the chemical variations within the tourmaline itself.

The authors are an all-star team that combines breadth and depth of technical expertise in the complementary skills of field collecting, mineral analysis, and geochemistry. Together, they tell the rich story of a mineral locality with a glorious past and a bright future for mineral collectors and earth scientists. From the detailed and precise history, to the stunning specimen photos, this book will be enjoyable and useful to collectors at all levels from beginner to veteran.

—Robert J. Lauf

ACKNOWLEDGMENTS

This book is based on an extensive database of field observations, specimens, and measurements built up over a period of fifty years. Although three of the authors (Chamberlain, Robinson, Walter) have spent countless hours collecting at the locality, beginning in the 1960s, our interactions with many other collectors have also greatly added to our base of knowledge about this complex locality. In the past decade, expert collecting by Scott Wallace and Donald M. Carlin Jr., in particular, have greatly enriched our understanding of the locality. We are especially grateful to the owners, Bower Powers Sr. and Bower Powers Jr., for granting continuous access to the site throughout this period.

We specifically want to acknowledge special assistance from our external Schiffer editor, Dr. Robert Lauf, and our internal Schiffer editor, Cheryl Weber. Michael Hawkins, collections manager at the New York State Museum, was an invaluable source of assistance. Ken Bart at Hamilton College kept the SEM/EDS system running flawlessly for our frequent use. Susan Robinson provided expert advice on the images and Helen Chamberlain edited and proofread the text.

The three authors for whom this locality has long been a passion are particularly grateful that the other three authors (Bailey, Chiarenzelli, Lupulescu) were infected by our enthusiasm and brought their specific areas of expertise to bear on critical aspects of this study. Without having the team resources represented by the six authors, many questions could not have been answered, discoveries could not have been made; and a coherent picture of this famous, prolific locality might not have emerged.

Photo Credits and Artwork. Photographs are credited by name in the captions except for the authors, which are credited as: SCC (Steven C. Chamberlain), GWR (George W. Robinson), MW (Michael R. Walter), JC (Jeffrey R. Chiarenzelli), ML (Marian V. Lupulescu), and DGB (David G. Bailey). Note that all sizes are maximum dimensions. Artwork is credited in the captions. Uncredited maps and diagrams were produced by the authors.

Collectors and Dealers. Where known, the names of those who collected the illustrated specimens, and when, have been given. Much of the data has been archived over generations, going back to the discovery of the locality in 1859. The efforts of past collectors and mineral dealers to preserve such information and keep it with the specimens as they were distributed and redistributed have greatly enhanced their value and demonstrate the interface between history and science.

INTRODUCTION

Figure 1. Watercolor of a black tourmaline specimen from Pierrepont. *Artist Brandi Zzyzx*

Black tourmaline crystals from Pierrepont in St. Lawrence County, New York, are familiar to most collectors. Indeed, many contemporary field collectors have visited the occurrence because it has been open as a fee locality for seventy-five years. Most nineteenth-century collections also contain specimens of Pierrepont black tourmaline since collecting occurred intermittently from its discovery in 1859 until the beginning of the twentieth century. Moreover, huge numbers of collector-quality specimens have been retrieved. The combination of the widespread familiarity of the specimens, their extensive representation in worldwide institutional collections, the status of the occurrence as a productive locality for nearly 150 years, and the continuing availability of the site for further specimen production makes Pierrepont an important historic, classic, and contemporary mineral occurrence. There appears to be virtually nothing in the published specimen mineralogy literature about this classic mineral locality beyond mentions in field-trip travelogues or its inclusion in various summaries of the minerals found in northern New York. The absence of definitive published

Figure 2. Acrylic painting of a black tourmaline crystal cluster from Pierrepont. *Artist Susan Robinson*

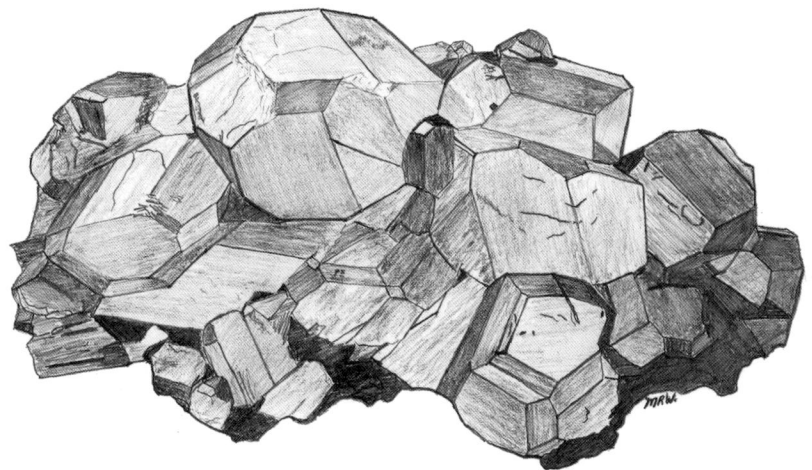

Figure 3. Pencil sketch of a black tourmaline specimen from Pierrepont. *Artist Michael R. Walter*

information was the initial motivation for this book. As we proceeded, we came to understand just how much heretofore unrecognized information about history, geology, and specimen mineralogy was tucked away in published sources. What follows is our attempt to present comprehensive information up to the frontier of what is known. We also raise some of the more important or interesting questions coming from our investigations that we could not answer without further research. In this way, we satisfy the needs of the mineral collector to know what occurs at the locality, what it looks like, and how it formed, as well as the needs of future researchers who wish to begin at the cutting edge of what is already known.

For more than a century, the focus was on the black tourmaline—the accessory minerals, though mentioned, were clearly deemed of secondary importance. Because unstriated, brilliant, nearly equant black tourmaline crystals of the sort found at Pierrepont are not found in any quantity elsewhere, serious collectors and researchers implicitly understood that there was something unusual, and probably complex, about this locality. The discovery of the smoky quartz vein in the early 1970s reawakened some interest in the details of the site's geology and mineralogy. However, it was the further discovery and systematic excavation of the four streamside veins in the twenty-first century that catalyzed serious study and brought the accessory minerals out of the shadows.

We address the history of the locality in chapter 1. We now know who discovered the occurrence, in the sense of the first visit by a mineral collector, and when. We know the complete history of land ownership from the Macomb Purchase in 1792 to the present day. We have a much better understanding of the geological origin of the deposit—it originally crystallized from a liquid—although many details remain to be investigated as discussed in chapter 2. We present a reasonably comprehensive description of the eight distinct collecting sites that have been found since the locality's discovery and provide mineral descriptions for each in chapter 3. In chapter 4, we show a gallery of black tourmalines in museum collections and highlight additional exceptional specimens of black tourmaline not previously included as figures. Finally, we address, to varying degrees, special topics of interest in chapter 5, including what the exact composition of the black tourmaline and what its proper name should be, the large number of pseudomorphs present at the occurrence, and the realization that some minerals, especially rare-earth species, formed much later than the original Precambrian pegmatite.

The six authors represent the kind of multidisciplinary expertise necessary to address complex topics like the black tourmaline occurrence near Pierrepont, New York. Although different authors have contributed more specifically to some parts than others, the end result is blended and enriched by the mingling of contributions from all the authors. Hence, we have a uniform presentation and organization throughout without listing authors for specific chapters. George Robinson, Steven Chamberlain, and Michael Walter provided the fundamental firsthand information on field collecting. George Robinson, ably assisted by Susan Robinson, Michael Walter, and Steven Chamberlain, assembled the historical record. Jeff Chiarenzelli, George Robinson, Marian Lupulescu, and David Bailey hammered out the geological story. Marian Lupulescu focused on the composition of the black tourmaline, which turned out to be more complicated than initially suspected and also identified interesting trace minerals in thin section. Steven Chamberlain and David Bailey provided modern analyses of many of the mineral species using SEM/EDS (a semi-quantitative scanning electron microscope). George Robinson and Jeff Chiarenzelli performed analyses using XRD. George Robinson procured images and information about Pierrepont specimens in museums. All the authors wrote first drafts of various parts of the book; these were woven into the final narrative by a round-robin process of editing and revision. The photographs are individually credited, but most were taken by the authors.

We hope this volume will prove fun to look at and interesting to read, and that it just might trigger further interest in this extraordinary mineral locality.

CHAPTER 1

HISTORY

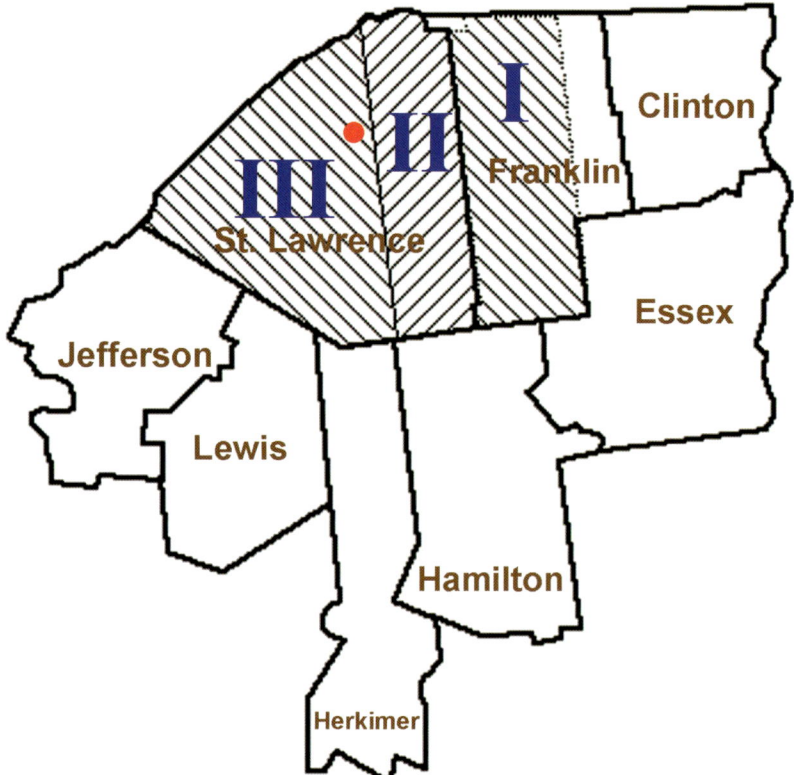

Figure 4. Map of counties in northern New York showing the first three great tracts of the Macomb Purchase. The black tourmaline locality, shown by the red dot, is in tract III.

Ownership History

After the colonial period, the land encompassing the black tourmaline locality was owned by New York State. In 1792, efforts to avoid state bankruptcy included an enormous land sale of almost 4 million acres in northern New York. This became known as Macomb's Purchase, since the official buyer was Alexander Macomb (1748–1831). In fact, a triumvirate of Macomb, William Constable (1752–1803), and Daniel McCormack (1739/40–1834) made the purchase. Constable's part of the deal was a section of northern St. Lawrence and Franklin Counties, including the present town of Pierrepont.

Constable's business went bankrupt. However, his prosperous son-in-law, Hezekiah Beers Pierrepont (1790–1838), acquired portions of Constable's land holdings, including those in the town of Pierrepont in St. Lawrence County. Ultimately both the town of Pierrepont and village of Pierrepont were named after him. After his death, Pierrepont's sons, Henry E. Pierrepont and William C. Pierrepont, managed the land holdings as trustees of the estate, successfully subdividing them into small parcels and selling them off.

On December 1, 1852, farmer Stephen A. Crary purchased seventy-one acres from the Pierrepont estate, and an additional twenty-four acres on December 1, 1859. Upon his death in 1880, Stephen Crary bequeathed his property to his son, Ryland A. Crary. On May 26, 1894, John Holcomb purchased the property from Ryland Crary and later sold it to Bower Powers Sr. on February 3, 1939. Bower Powers Sr. increased the size of the farm with additional land purchases in 1939 and 1947. Bower Powers Jr., purchased the farm from his father on September 4, 1966, and still owns the property.

To summarize, the post-colonial ownership of the black tourmaline locality northwest of the village of Pierrepont was:

New York State   pre 1792
William Kieran Constable   1792–c. 1801
Hezekiah Beers Pierrepont   c. 1801–1838
Hezekiah Beers Pierrepont estate   1838–1852
Stephen A. Crary   1852–1880
Ryland A. Crary   1880–1894
John Holcomb   1894–1939
Bower Powers Sr.   1939–1966
Bower Powers Jr.   1966–present

Ownership history was compiled from public access records at the county seat of St. Lawrence County in Canton, New York, and open access files in the collections of the New York State Library (Pierrepont Family Papers) and the New York Public Library (Constable-Pierrepont Papers).

Collector History

Discovery—Roselle Theodore Cross. The mineralogical discovery of the black tourmaline locality along Leonard Brook near Matthew's saw mill was made in 1859 by a teenager from Richville, New York. Roselle Theodore Cross was fifteen or sixteen when a local farmer showed him the locality. He first described his discovery in a letter to the editor of *The Mineral Collector* in 1899, responding to an article by W. S. Valiant, the curator at Rutgers University, about collecting in St. Lawrence County, New York. In part, Cross wrote:

> About forty years ago, while attending a religious meeting with my father, I found the now famous locality for black tourmaline in Pierrepont, St. Lawrence Co. Through some specimens that I gave to a friend Mr. Nims got on the track of it, hunted it up, exploited it for years, and has sold many a wagon-load of beautiful black crystals from that wonderful half acre. I have been there several times, and the last time, although the locality was said to be exhausted, I brought away several hundred specimens.—R. T. Cross, York, Nebraska, March 1899. (*The Mineral Collector*, vol. 6 (2) [April 1899] 33–34)

Figure 5. Roselle Theodore Cross in 1870. *Courtesy of the Oberlin College Archives*

R. T. Cross was born in Richville in St. Lawrence County in 1844. He graduated from Oberlin College, studied theology at Union and then Andover Seminary, and was ordained as a minister in the Congregational Council in 1869. He was a mineral collector from an early age and continued this activity throughout his long ministerial career. He died in 1924 in Twinsburg, Ohio, having written a number of autobiographical books. In *Crystals and Gold*, published in 1903, he gives a more extended account of his discovery of the black tourmaline locality:

> My father took me with him once to a religious meeting twenty-five miles from home, up on the hills of Pierrepont. At the farmer's house where we stopped I saw a shining black crystal of tourmaline. In reply to my inquiry as to where it came from, they said it was found in abundance near an old saw mill about half a

mile distant. The next morning they took us to the place. I dug the black brilliants for a half hour or so and then a thunder storm drove us away. I remembered the place for years and often wished that I could revisit it. I did so in 1871, while home on a vacation from teaching in Oberlin College. The crystals were remarkably smooth and brilliant and were found in abundance by digging in the dirt and decomposed rock. I gave some to a farmer in a distant part of the county. Mr. Nims, referred to above, saw them and followed up the clue until he found the locality, and from that place also he sold wagon loads of tourmaline crystals. It has been a famous locality for no blacker, or more brilliant, or more sharply cut crystals of tourmaline are found in this world. I visited the place in 1871, 1875, 1883, and 1890, and always came away with hundreds of specimens, good, bad, and indifferent. I always greatly enjoyed cleaning the Pierrepont tourmalines. Water beautifully brings out their bright blackness, and if there are dingy spots on the crystals they re-appear very slowly after the crystals are washed. It was almost as much fun to clean them as to dig them. More than once I have taken back into my collection specimens that I had discarded, so changed was their aspect after the rain had fallen on them.

By rights, the earliest name for the locality should be the Stephen A. Crary farm, Pierrepont, St. Lawrence Co., New York. However, that name seems never to have been published, and we are aware of no labels from the early period indicating this land ownership. Instead, his son, Ryland A. Crary, is usually listed in the original name of the locality—Ryland A. Crary farm—probably because he was the owner when the locality first appeared in the *Gazetteer of American Mineral Localities* as an appendix to *Dana's System of Mineralogy* (6th edition) in 1892.

Specimens distributed worldwide—Charles (Chester) D. Nims. Although Cross discovered the black tourmaline locality, it was undoubtedly Charles (Chester) D. Nims who put it on the map. Nims was born in 1811 in nearby Jefferson County and pursued farming and lumbering for thirty years. After a two-year successful sojourn in California, he returned in about 1850 with a serious interest in minerals. He built a business as a field-collecting mineral dealer and became a leading source of minerals from northern New York, and probably *the* leading source of black tourmaline from Pierrepont. Specimens acquired from Nims in the thirty-year period, 1860 to 1890, can be found in major collections around the globe. Nims listed his occupation as mineralogist in both the 1870 census and the 1880 census and was successful enough to move from a modest house on Sand Street in Philadelphia, New York, to a much grander Victorian house on Sand Street around the curve from his first house.

The catalogs of both the Oren Root Collection and the John H. Caswell Collection indicate the purchase of many specimens from Nims, including numerous specimens of black tourmaline. A detailed analysis of the Caswell

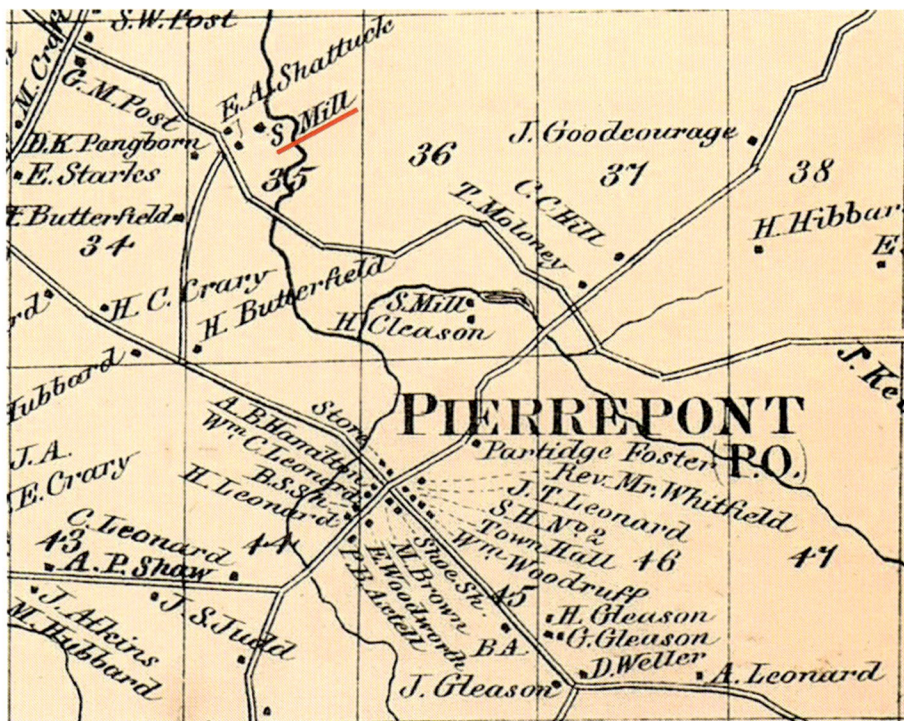

Figure 6. Matthew's saw mill, underlined in red, was across the creek from the locality in this 1865 atlas (Beers & Beers).

Chapter 1: History

Figure 7. C. D. Nims' second residence on Sand Street in Philadelphia, New York. Is this the house that black tourmaline built? *SCC*

catalog showed that of 3,712 entries, 299 specimens came from Nims, and of these, forty-nine were black tourmalines from Pierrepont. The tourmalines were acquired between 1878 and 1890. Moreover, only two other black tourmalines from Pierrepont appear in the catalog: one from Albert Chester and one from Oren Root, both acquired in 1882 and both with unusual crystal faces. It is likely that they were originally from Nims also. One wonders whether he might have had an exclusive collecting arrangement with Stephen A. Crary and then his son, Ryland A. Crary.

Fortunately, copies have been preserved of Nims' unpublished letter dated May 17, 1886, and sent to Professor Albert H. Chester. In it, Nims describes numerous northern New York localities that he exploited for his business, including the black tourmaline locality at Pierrepont:

> One mile north of Pierrepoint Centre near Matthew's saw mill, on the land of Uria Crary, Black Tourmaline

Figure 8. Cluster of black tourmaline collected by Nims about 1880. From the John I. Legro collection. 4.5 cm. SCC#8110. *SCC*

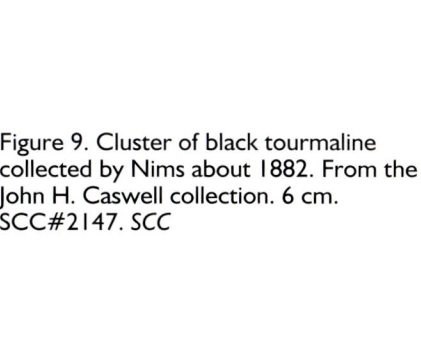

Figure 9. Cluster of black tourmaline collected by Nims about 1882. From the John H. Caswell collection. 6 cm. SCC#2147. *SCC*

11

Figure 10. Black tourmaline and quartz collected by Nims. From the Charles Upham Shepard collection. 12 cm. SCC#15435. *SCC*

heretofore found very abundant in groups with frequently an outlier doubly terminated and very nearly perfect, thought by some to be the best Tourmaline locality in the world, they occur with Quartz, Pyroxene and Mica finely crystallized, of a brilliant glossy black, are now very scarce and hard to get. (Nims 1886 letter, 7)

Despite investigating the local genealogical records of the Crary family, the identity of Uria Crary is not established. It may well have been a nickname for either Stephen A. Crary, or more likely his son, Ryland A. Crary, who owned the land at the time Nims wrote his letter.

The first crystal drawings of tourmaline from this locality were published by R. H. Solly in 1884 and likely described specimens Nims collected. His paper concerned crystallography and was based on specimens in the collection of Cambridge University, although he mentions examining Pierrepont tourmaline specimens from several other collections, presumably European. This suggests how widespread the distribution of these black tourmaline specimens had already become.

Twentieth century. The first published photograph of a specimen of tourmaline from Pierrepont appeared in 1901 as the frontispiece of vol. 8, no. 10 of *The Mineral Collector*. The specimen of quartz and black tourmaline was then in the collection of Frank B. Jones of Westfield, New Jersey.

In 1921, William M. Agar published a firsthand account of the top of the hill area:

This is another famous collecting ground. It is doubtless the locality from which most of the black tourmaline in the mineral collections of the country has come. It is situated on the right bank of Leonard Brook 1.8 km. (1.1 m.) northwest of Pierrepont Crossroads, ½ km

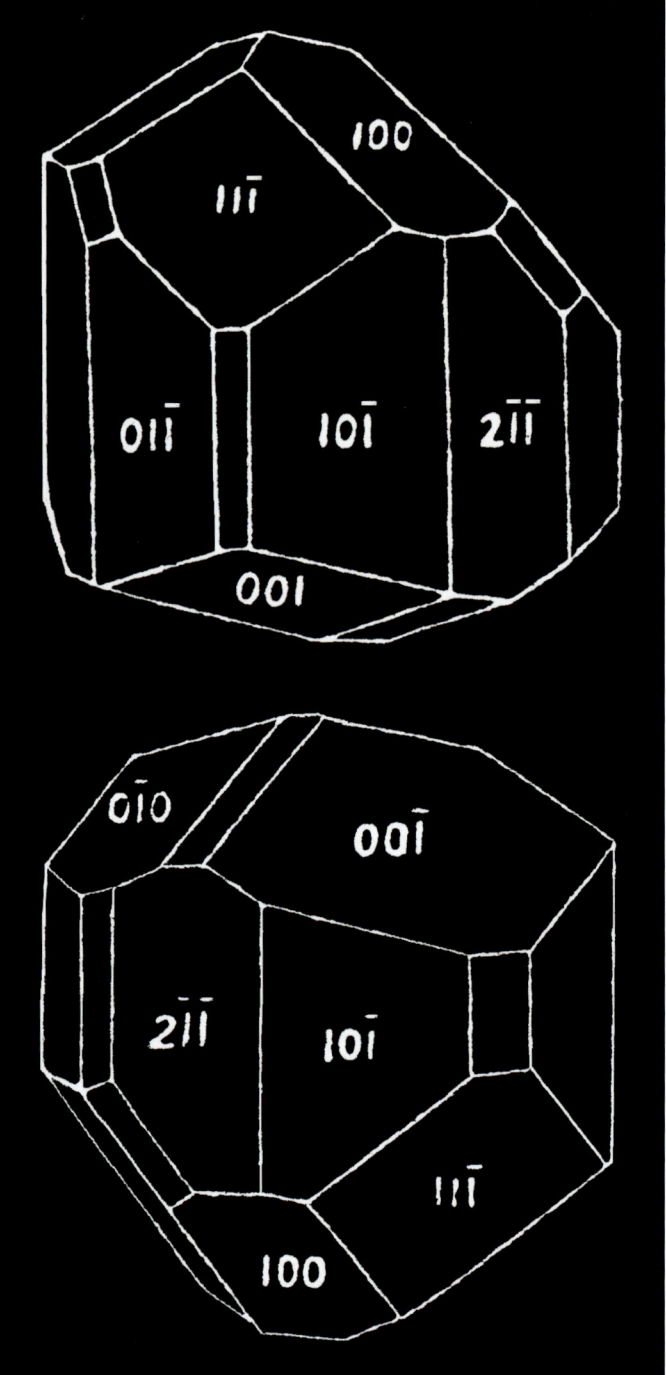

Figure 11. The first published crystal drawings of tourmaline from Pierrepont (Solly, 1884).

(0.3 m.) downstream from the bridge at B.M. 597 on the road running southeast from Crary Mills and off the sheet 1.8 km. (1.1 m.) northeast of Pierrepont. The tourmaline occurs as a band running from the brook intermittently up the hill for about 150 meters. A great many pits have been blasted in it, but it still forms a very conspicuous black band on the slope. Clusters of brilliant black crystals are abundant and doubly

Figure 12. First published photograph of black tourmaline and quartz from Pierrepont (1901). From the Frank B. Jones collection. *George E. Ashby photo*

terminated, stubby, polar crystals can with care be dug out. They occur with quartz, some calcite, phlogopite, and pyroxene in good square crystals. These grow more abundant as the band is followed up the hill. (Agar 1921, 161)

This account indicates that Nims' hints that the locality was becoming exhausted were premature, and that the top of the hill was almost certainly the original site he exploited in the second half of the nineteenth century.

During the first half of the twentieth century, collecting seems to have been sparse compared to the large quantities of specimens C. D. Nims produced. We have not found a single specimen collected during the period 1894 to 1939 listing the locality as the John Holcomb farm. It may be that this locality, like so many in St. Lawrence County, was fallow and rarely visited—even somewhat lost. Alternately, it might be that John Holcomb did not allow mineral collecting on his property.

In 1948, Horace W. Slocum of Rock Hill, South Carolina, published a multipart article, "Rambles in a Collector's Paradise," in *Rocks & Minerals*. In these articles, he describes his collecting experiences across northern New York. In part 4, which appeared in the September/October issue, he describes his visit to the black tourmaline locality at Pierrepont. This article contains what seems to be the first photograph of the locality ever published. Slocum set out to find the band of tourmaline running up the hill as described by Agar in 1921. He did not, however, go far enough along Leonard Brook. Instead he found a boulder containing black tourmaline in a pasture along the stream, and collected there. After a group of twenty-three collectors visited this specific site, he declared that the famous black tourmaline locality was actually a glacial boulder and was now essentially exhausted:

I collected some good clusters on my first visit. Some fine clusters of square terminated pyroxene crystals as well. That boulder must have been a mass of pockets. The second visit was of course not so successful. About all I got then was experience.

If this is organized collecting spare me the consequences, for at this rate there'll be no more minerals left to collect in another fifty years.

Now even if this locality is pretty well cleaned up I am going to give directions for finding it. Some day some one may wish to go there and weep at its destruction. It has been a truly great locality in its time. (Slocum 1948, 772–773)

In 1948, when Slocum visited, a pasture ran from the access road (Post Road) along Leonard Brook almost to the original locality. Most of this large field is now wooded along the path into the locality and there are indeed many displaced boulders a short distance from the top of the hill.

Figure 13. "... this locality is pretty well cleaned up ..." (H. W. Slocum in 1948, Rocks & Minerals 23:773).

In the second half of the twentieth century, the locality became an increasingly well-known fee locality and was visited frequently by local collectors, collecting tourists, and organized groups. New discoveries revealed that the mineralized area was significantly greater than just the original site on top of the hill. For example, Elmer Rowley and Ralph Lapham from Glens Falls, New York, did most of their collecting on the hillside southeast of the classic locality on top of the hill adjacent to Leonard Brook. Schuyler Alverson also recovered many good specimens.

In the 1950s, Ronald Waddell, a Syracuse, New York, collector, discovered the first of four streamside veins running roughly perpendicular into Leonard Brook south of the original site (Chamberlain 2007). The Waddell Vein was subsequently excavated in several phases and produced spectacular specimens of tourmaline and other minerals.

In the early 1970s, Nick Rochester, working at the original site, discovered a vein in an open fracture, apparently much younger in age than the tourmaline mineralization, and filled with quartz and calcite crystals that formed in open spaces. During this period, collectors also discovered that the wetlands (swamp) adjacent to the hillside site southeast of the original site contained collectible specimens. By the end of the century, there were multiple sites on the property where specimens could be collected, and some local collectors began to understand that much of the locality might still be unexplored. About this time, serious scientific studies of the mineralogy and geology of the occurrence also commenced. Local collectors began to pay closer attention to who was collecting specimens, what was found, and the site-specific records for specimens

Figure 14. Candid portrait of Ron Waddell in 1983 holding a blue fluorite from Penfield. *Fred Ruhe photo*

Figure 15. Black tourmaline collected by Ron Waddell in 1964 from the Waddell vein. 12.5 cm. SCC#12151. *SCC*

from the locality. Accessory minerals gained more importance as questions of origin and paragenesis were considered.

Twenty-first century. Three of the authors (George Robinson, Steve Chamberlain, and Michael Walter) have intermittently done extensive collecting at various sites on the property since at least the 1960s. In the early 1960s, George Robinson began collecting in the boulder field in what was then a cow pasture, but later concentrated on more productive sites across the main collecting area on top of the hill. Frequent visits from 1964 to 1970, often in the company of Schuyler Alverson, Robert Dow, and Richard Hansen, yielded numerous fine specimens of tourmaline with uralite, the best of which are now in the collection of the Canadian Museum of Nature, Ottawa, Ontario. All three tended to retain and preserve unusual and scientifically important specimens in addition to the spectacular black tourmaline. The search for a broad range of specimens intensified in the twenty-first century, and Michael Walter became perhaps the most active field collector at the site, logging more than 300 collecting days since 2000. Steve Chamberlain concentrated his efforts on building an extensive suite of specimens from the locality, which will be preserved at the New York State Museum.

In spring 2004, Scott Wallace and Michael Walter began to extend the small, forty-year-old trench on the Waddell Vein. When they began, it had been dug no deeper than a meter. A seasonal lease with the landowner gave them exclusive collecting rights at this site. By winter they had collected many noteworthy specimens of tourmaline and other minerals. Their collecting continued through the 2005 season, but with less success. By the end of 2005, the length of the trench on the Waddell Vein had reached its current size and depth and was below the water table, requiring mechanical pumping to keep it dewatered. Wallace continued to work the west wall intermittently during each collecting season from 2006 to 2008, recovering excellent specimens of some accessory minerals.

In spring 2008 with another seasonal lease, Michael and Jay Walter began excavation of a newly discovered vein that ran roughly parallel to the Waddell Vein about twenty meters to the south. A season of steady excavation yielded many unusual and interesting specimens in addition to tourmaline. Michael Walter kept a detailed journal of when and where specimens were recovered. Noteworthy fluorapatite crystals, unusual pseudomorphs, and late-stage, rare-earth minerals were collected and documented. The fluorapatite crystals that were recovered greatly exceeded the norm in both size and quality. Many various pseudomorphs were recovered here, including the first known quartz pseudomorphs after phlogopite found anywhere. The first synchysite-(Ce) from the property was also noted here. A summary of the mineralogy (Chamberlain, Bailey, Walter, 2010) and a history of the streamside veins (Walter and Chamberlain, 2010) were published. The abundance of fluorapatite in this vein led to the informal name Phosphate Vein.

Toward the end of the 2008 collecting season, Michael Walter excavated a small vein midway between the Waddell Vein and the Phosphate Vein, hence its name of Middle Vein. The most important displayable specimens found included some very elongated mica crystals.

Figure 16. Scott Wallace in the Waddell Trench in 2008. SCC

Figure 17. Black tourmaline collected by Scott Wallace from the Waddell vein. 5 cm. SCC#18292. SCC

Figure 18. Mike Walter working at the bottom of the Phosphate Trench, Summer 2008. *Barbara Walter photo*

Figure 19. Black tourmaline from the Phosphate Vein collected by Mike Walter in the summer of 2008. 10.8 cm. MW#00089. *MW*

Figure 20. Donald Carlin Jr. digging on top of the hill in 2008. *SCC*

By the end of summer 2008, Donald (Donnie) M. Carlin Jr. had joined Scott Wallace in the Waddell Trench looking for lateral offshoots of the main vein, without much success. In September, they moved their efforts several meters north and found a fourth parallel vein, the Wallace-Carlin Vein. This site had been pointed out by a visiting friend of Wallace's who found some promising tourmaline near the surface. Excavations that fall and the following spring and summer opened another productive vein similar to the adjacent Waddell Vein, but with more goethite than the other three streamside veins and notable quantities of marcasite.

From 2010 through 2013, Donnie Carlin extensively worked at the original site on top of the hill. He encountered the smoky quartz vein at depth and retrieved many specimens. He also recovered some large, very sharp mica crystals from the marble adjacent to the vein.

For much of the collecting season in 2014, Michael and Jay Walter worked a vein rich in uralite (tremolite ps. diopside) along the northern edge of the original locality. They recovered large quantities of this material, some in combination with tourmaline. They also found numerous other species uncommon to the locality, including magnetite, pyrite, actinolite, and tremolite.

The past decade of collecting activity has produced large quantities of tourmaline specimens that rival those found here in the past. Moreover, the past decade has yielded a much wider range of excellent specimens than in the locality's previous history.

Gallery. We close this chapter with a gallery of fine specimens extracted by other serious collectors in the second half of the twentieth century. Each of these collectors spent many days mining at the original site on top of the hill. They include: teenager William Pinch from Rochester, New York; Schuyler Alverson from Rensselaer Falls, New York; SUNY Potsdam college student George Robinson from Glens Falls, New York; Ivan McIntosh from Gouverneur, New York; collector-dealers Bill and Sonja

Figure 21. Black tourmaline collected by Donald Carlin Jr. from the Wallace-Carlin Vein in 2008. 8 cm. SCC#9468. *SCC*

Bieler from New Jersey; William S. Condon from Syracuse, New York; Steven Chamberlain from Manlius, New York; professional miner and collector Vernon Phillips from Spragueville, New York; professional miner and collector Terry Holmes from Edwards, New York; James Dee from Alden, New York; professional miner and collector, Charles Bowman from Gouverneur, New York; and collector-dealer Eric Greene from Greenfield, Massachusetts. This listing is by no means comprehensive; rather it illustrates the breadth of the serious involvement of members of the collector community in procuring excellent specimens from this locality.

Figure 22. Black tourmaline collected by William Pinch in 1957. 9 cm. SCC#14709. *SCC*

Figure 23. Black tourmaline collected by Schuyler Alverson in 1965. 5.1 cm. NYSM#21048. *Stephen Nightingale photo*

Figure 24. Black tourmaline and phlogopite collected by George Robinson in 1968. 6.5 cm. SCC#6383. *SCC*

Figure 25. Black tourmaline floater collected by Ivan McIntosh in 1971. 4.5 cm. SCC#6383. *SCC*

Figure 26. Black tourmaline collected by Bill & Sonja Bieler in 1980. 7 cm. SCC#6760. *SCC*

Figure 27. Black tourmaline collected by William S. Condon in 1981. 6.6 cm. SCC#12519. *SCC*

Figure 29. Black tourmaline collected by Steve Chamberlain in 1985. 7.5 cm. SCC#6445. *SCC*

Figure 28. Black tourmaline collected by Vernon Phillips in 1983. 7 cm. SCC#931. *SCC*

Figure 30. Tabular black tourmaline collected by Terry Holmes in 1987. 8 cm. SCC#11050. *SCC*

Figure 31. Black tourmaline and quartz collected by James Dee in 1988. 6.2 cm. SCC#9118. *SCC*

Figure 32. Quartz on black tourmaline collected by Charlie Bowman in 1990. 9.8 cm. SCC#9129. *SCC*

Figure 33. Black tourmaline collected by Eric Greene in 1996. 6.2 cm. SCC#15570. *SCC*

LITERATURE CITED

Agar, W. M. "The minerals of St. Lawrence, Jefferson, and Lewis Counties, New York." *American Mineralogist* 6 (11) (1921): 148–164.

Beers, S. N. and D. B. Beers. *New Topographical Atlas of St. Lawrence Co., New York.* Philadelphia: Stone & Stewart, 1865.

Chamberlain, A. "Tourmaline and quartz, St. Lawrence Co., N. Y." *The Mineral Collector*, vol. viii, (10) (1901), frontispiece.

Chamberlain, S. C. "Excavation of a pegmatite dike on the Bower Powers farm, Pierrepont, St. Lawrence County, New York." *Rocks & Minerals* 82 (2007): 233–234.

Chamberlain, S. C., D. G. Bailey, and M. Walter. "Minerals of the phosphate vein, Bower Powers farm, Pierrepont, St. Lawrence County, New York." *Rocks & Minerals* 85 (2010):162–163.

Cross, R. T. "Anent, the Collector's Paradise." *The Mineral Collector*, vol. vi, no. 2 (1899): 33–34.

Cross, R. T. *Crystals and Gold.* York, Neb.: The Nebraska Newspaper Union, 1903.

McMartin, B. *The Great Forest of the Adirondacks.* Utica, NY: North Country Books, 1994.

Slocum, H. W. "Rambles in a collector's paradise, Part 4." *Rocks & Minerals* 23 (1948): 771–777.

Solly, R. H. "On the tetartohedral development of a crystal of tourmaline. *Mineralogical Magazine* 6 (1884): 80–82.

Walter, M. R. and S. C. Chamberlain. "History of the Powers farm streamside veins, St. Lawrence County, New York." *Rocks & Minerals* 85 (2010): 556.

CHAPTER 2

GEOLOGY

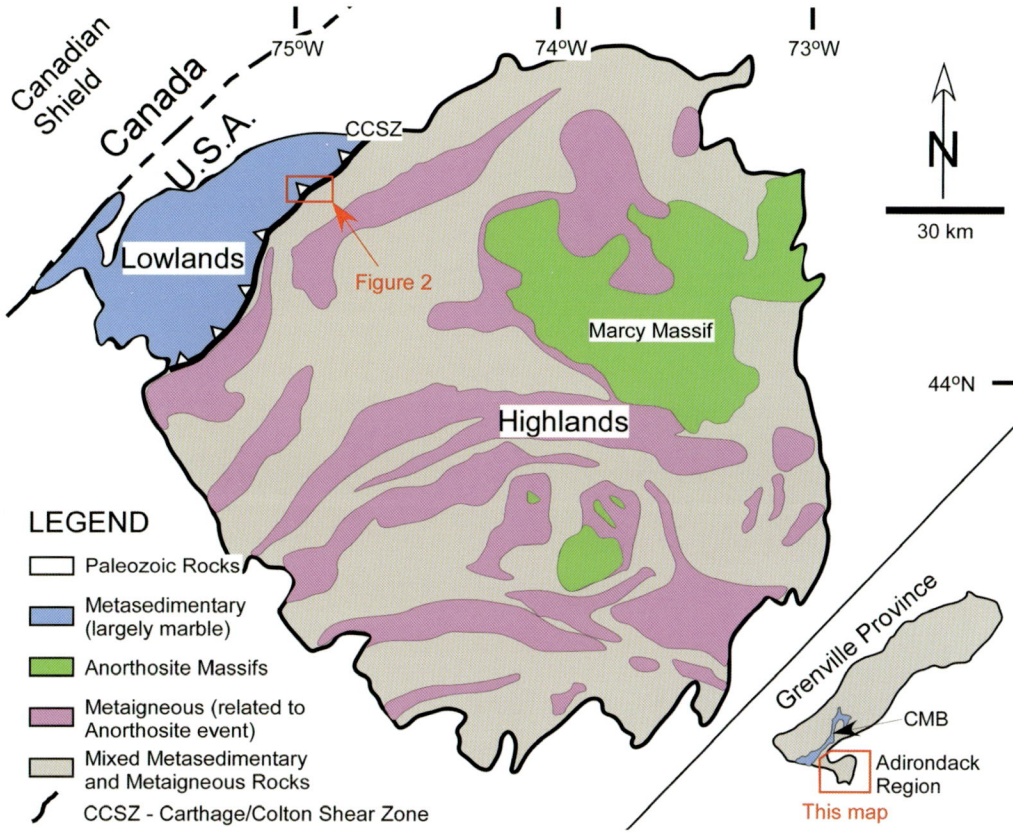

Figure 34. Simplified geological map with vicinity of the Pierrepont black tourmaline (Powers Farm) locality indicated by the upper red rectangle within the Adirondack Region. Inset on bottom right shows the location of the Adirondack Region within the Grenville Province. Note the location of the Central Metasedimentary Belt (CMB in insert) where both rocks and mineral deposits, similar to those in the Adirondack Lowlands, are found.

Introduction and Tectonic Setting

While the importance of the black tourmaline and associated minerals at the Powers Farm locality near Pierrepont is firmly established (e.g. Chamberlain and Robinson, 2013), relatively little is known about the origin of the deposit. In this chapter, we examine the geology of the area and constraints on the origin of the deposit, including its age and mode of formation. In doing so, it is important to view the deposit in its context as one of a large number of mineral occurrences throughout the Adirondack Region (fig. 34), particularly the Adirondack Lowlands.

The Adirondack region exposes Precambrian basement rocks that are part of the Canadian Shield, a vast, stable area that forms the nucleus of North America. It is also part of the Grenville Province, the exposed roots of an approximately billion year old mountain belt that once extended through Scandinavia, Greenland, the Canadian Maritimes, Quebec, southern Ontario, and the Adirondacks, as well as isolated basement inliers up and down the spine of the Appalachian Mountain Chain and beyond. The Grenville Province is named after the rocks exposed in

Grenville, Quebec (Logan, 1863) and is characterized by a series of orogenic (i.e. mountain-building) events between about 1,250 and 1,000 million years ago collectively known as the Grenville Orogeny (Rivers, 2008).

The Grenville Province represents a broad zone of highly deformed and metamorphosed rock that formed during the collision between ancestral North America and South America (McLelland et al., 2013). During this collision, one plate was partially subducted beneath another—similar to modern India and southern Asia—and the crust was thickened, and the rocks in the Adirondacks were forced downward in the crust to depths of up to 30 km. The resulting deformation and metamorphism included the growth of new minerals in equilibrium with temperatures and pressures deep in the crust and folding and the development of metamorphic fabrics. In addition, temperatures were hot enough to initiate melting in some rock types.

Following the Grenville Orogeny, a long period of erosion wore the mountains down to sea level. By 550 million years ago, basement rocks in the Adirondacks were buried by a thin layer of much younger, flat-lying, lower Paleozoic sandstones (including the Potsdam Sandstone), limestones, and shales. The impact of tectonic events that formed the younger Appalachian Mountains on basement rocks of the Adirondack Region was relatively minor as the new mountain chain rose along the eastern seaboard.

By 180 million years ago, the entire Adirondack Region began to uplift (Roden-Tice and Tice, 2005) and erosion began to strip away the overlying Paleozoic sedimentary rocks. In the Adirondacks, a large NNE-elongate, domal area exposed the basement rocks at the surface once again as the roots of an ancient mountain belt became new mountains. Today, the contact between the Potsdam sandstone and the Grenville basement rocks can be observed dipping gently away from the center of the dome in a radial pattern approximately near the location of the Adirondack Blueline (boundary of the Adirondack Park). In the Lowlands, it is not uncommon to find patches of the basal Potsdam sandstone lying directly on top of older Grenville basement rocks (fig. 35).

Geology of the Adirondack Region

The Adirondack Region is typically divided into the northwest Lowlands and Highlands, which although differing in several respects, share much of their geologic history. In the northwest Adirondack Lowlands, where Powers farm is located, rocks were last metamorphosed and deformed during the Shawinigan Orogeny (about 1,180 to 1,160 million years ago) (Heumann et al., 2006). This event led to the deformation of all rocks that existed at that time in parts of southern Ontario, Quebec, and New York. This area is collectively known as the Central Metasedimentary Belt (CMB in the inset of Fig. 34) because of the similarity in rock types, including a thick sequence of marble. This explains some of the many similarities in the mineral occurrences found in the Adirondacks and adjacent areas of the Grenville Province.

To the southeast of the Lowlands, the Adirondack Highlands consist primarily of metamorphosed igneous rocks intruded between 1,165 and 1,155 million years ago. These were last deformed during the Ottawan Orogeny, nearly 100 million years after the Shawinigan Orogeny in the Lowlands (about 1,050 million years ago). The boundary between the Highlands and Lowlands is known as the Carthage-Colton Shear (or Mylonite) Zone (C-CSZ) (Geraghty et al., 1981; Selleck et al., 2005). To explain the many differences between the Highlands and Lowlands,

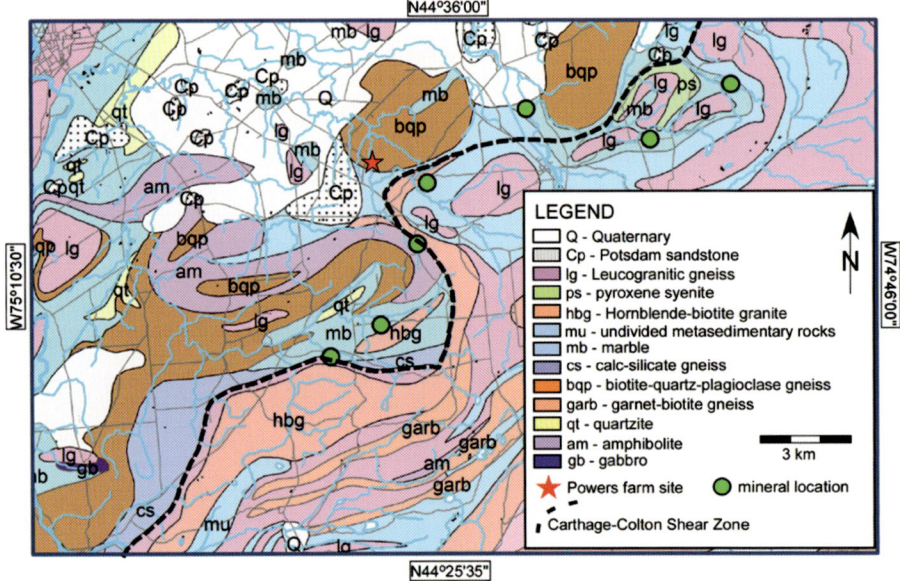

Figure 35. Geologic map of the Powers farm location (red star) and surrounding area (after Isachsen and Fisher, 1970). Green circles indicate other known, nearby calc-silicate mineral localities. The interpreted trace of the Carthage-Colton Shear Zone (C-CSZ), the boundary between the Adirondack Highlands and Lowlands, is shown with black dashes.

investigators have proposed that the C-CSZ is a long-lived fault with both an early ductile, and late brittle, history (Streepey et al., 2001). The Lowlands are believed to have dropped down off the Adirondack Highlands during extensional faulting associated with the cessation of mountain building and the relaxation of the crust (Selleck et al., 2005).

Although both regions consist of rocks that were once deep in the crust, the C-CSZ marks a distinct change in metamorphic grade and the abundance of metasedimentary rocks (fig. 34). The overall pattern of the many radiometric ages determined in the past several decades further confirm this fundamental boundary (Heumann et al., 2006). These features have resulted in the many differences in the mineral occurrences in each region.

Generations of geologic mapping and zinc exploration have led to structural and stratigraphic models for the Adirondack Lowlands. Currently the stratigraphy is believed to comprise three major lithologic units consisting of, bottom to top: the Lower Marble, the Popple Hill Gneiss, and the Upper Marble—all part of the marble-rich Grenville Supergroup (e.g. Lucas and St-Onge, 1998). These units have been affected by a number of NE-trending folds and are overturned to the southeast, much like stacked and crumpled slices. The most detailed stratigraphic subdivision comes from the Sylvia Lake Syncline. Here exploration for zinc has established a stratigraphic sequence consisting of sixteen subunits, which together comprise the Upper Marble (Brown and Engel, 1956). Several igneous rock suites ranging in age from 1,200 to 1,150 million years ago (Peck et al., 2013) cross-cut the metasedimentary rocks and form numerous sill-like layers or elongate, oval, bodies.

Detailed Geology of the Powers Farm Locality

Powers farm is just within the Adirondack Lowlands near the C-CSZ. On the Adirondack Sheet (Isachsen and Fisher, 1970) of the New York State geological map, the area is underlain by biotite-quartz-plagioclase gneiss (bqp) known as the Popple Hill Gneiss and a unit (mu) that represents undivided metasedimentary rock (figs. 35 & 36). To the south and east, directly across the C-CSZ from the Powers farm, is biotite and/or hornblende granitic gneiss (hbg) considered to be part of the Highlands (fig. 35).

Detailed mapping of the Powers farm deposit (fig. 36) reveals that a wide variety of metasedimentary rocks occurs within a fairly limited area. These include garnet- and sillimanite-bearing biotite-quartz-feldspar gneisses, rusty pyroxene-bearing calc-silicate gneisses, and relatively pure marble, as well as various skarn-like assemblages and tourmaline-bearing rocks (fig. 37). Other rocks such as pegmatite, amphibolite, and quartzofeldspathic gneisses of variable composition are likely igneous in origin. As a group, they represent the typical diversity of rocks found

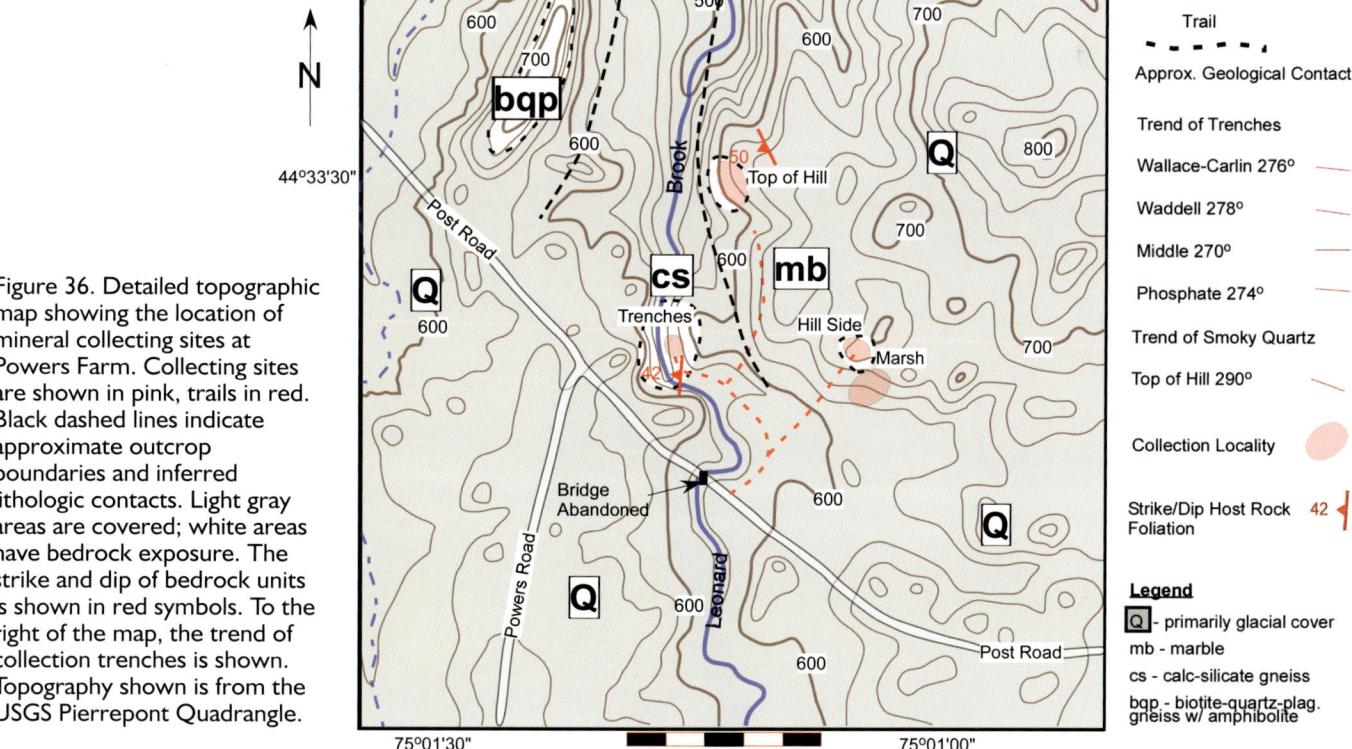

Figure 36. Detailed topographic map showing the location of mineral collecting sites at Powers Farm. Collecting sites are shown in pink, trails in red. Black dashed lines indicate approximate outcrop boundaries and inferred lithologic contacts. Light gray areas are covered; white areas have bedrock exposure. The strike and dip of bedrock units is shown in red symbols. To the right of the map, the trend of collection trenches is shown. Topography shown is from the USGS Pierrepont Quadrangle.

in the Adirondack Lowlands and display few, if any, characteristics that indicate a unique or different history from the surrounding area.

The variability in rock types across small lateral distances and the general lack of bedrock exposures in the immediate area (fig. 36) makes it difficult to produce a detailed map. However, rock units strike consistently south-southeast and dip moderately to the north-northwest. Given the thinness of the rock units and their state of relatively high ductile strain, it is not currently possible to assign them to any particular stratigraphic unit. A more likely scenario is that they represent the strong deformation inherent along the C-CSZ. Given the complicated nature of the C-CSZ (Geraghty et al., 1981; Streepey et al., 2001) and its meandering trace in this area, it is most likely not a simple planar boundary (fig. 35).

The Powers farm location, although unique in its abundance and quality of tourmaline, is but one of several known mineral occurrences along the approximate trace of the C-CSZ (fig. 35). This zone contains a large volume of sulfide-rich, rusty, diopside-bearing calc-silicate rock. The localities share a number of other characteristics, including development within, and adjacent to, similar rock types, proximity to marble and granitic pegmatite, and exceptionally large, well-formed crystals of calc-silicate minerals (often in direct contact with coarse calcite). Mineralized zones including tourmaline invariably cross-cut the surrounding host rocks, which often have a well-defined metamorphic fabric or foliation. This is an important observation because it indicates that the mineralization formed after the widespread and intense regional deformation associated with the Shawinigan Orogeny. It is also important to note that while not all mineral occurrences in St. Lawrence County occur within this zone or as close to the C-CSZ as the Powers farm, many do, and this may well be a clue to their origin.

At Powers Farm, the intrusive nature of the tourmaline deposits can be seen in a set of four parallel veins, or more accurately, dikes (Chamberlain, 2007) that have yielded fine mineral specimens along the bank of Leonard Brook (fig. 38). Although the mineral assemblage in each of the dikes varies, all four share a common orientation (E-W) and cut nearly perpendicular to the strike of the host gneisses exposed in the trenches (south-southwest). Each dike has been widened to permit access; however, they were generally 10 to 25 cm wide and up to 1 meter wide in parts of the Waddell Vein, and all had calcite cores. They have each been excavated almost to stream level, as much as 2 meters below the top of the bedrock. In the walls of each trench, the foliation of the quartzofeldspathic host gneisses dips west at nearly 30 degrees and is clearly cross-cut by the nearly vertical tourmaline-rich dikes (see

Figure 37. Photomicrographs of thin sections showing some of the lithologies found at Powers farm including: A) rusty, weathering, calc-silicate gneiss displaying both orthopyroxene and diopside; B) tourmaline and biotite-bearing quartzofeldspathic gneiss; C) pyrite and talc-bearing marble; and D) garnet, sillimanite, and biotite-bearing gneiss. Labels: am - amphibolite; bi - biotite; cal - calcite; di - diopside; grt - garnet; kfs - potassium feldspar; op - opaque; pl - plagioclase; py - pyrite; qtz - quartz; sil - sillimanite; tlc - talc. The technical details of each photograph are shown below it.

A. Calc-silicate gneiss, 40x magnification, crossed polarizers.

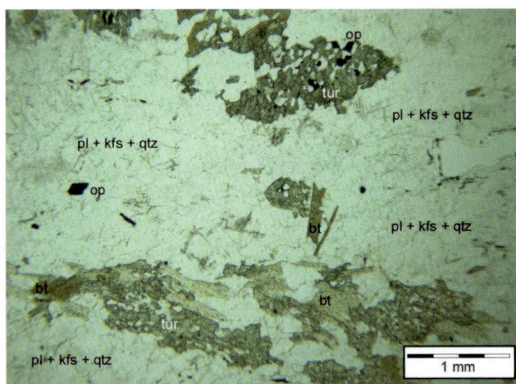
B. Quartzofeldspathic gneiss, 40x magnification, plane polarized light.

C. Pelitic gneiss, 40x magnification, crossed polarizers.

D. Marble, 40x magnification, crossed polarizers.

Figure 38. Photograph of one of the four parallel collecting trenches worked near the bank of Leonard Brook. The trench is nearly vertical and trends 278 degrees. The foliation in the bedrock can be seen dipping at a shallow angle to the left and away from the observer. The inset in the upper left corner shows the trace (apparent dip) of the foliation on the wall of the trend. Note rock hammer for scale.

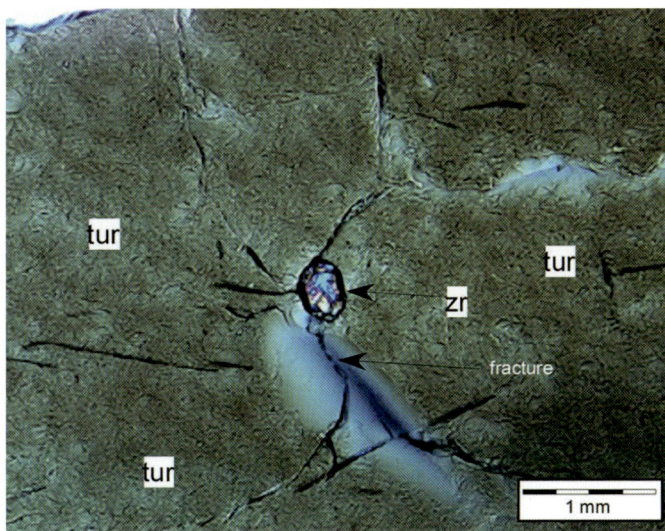

Figure 39. Photomicrograph of a zircon crystal included within tourmaline from Powers Farm. Note the sharp grain boundaries and euhedral shape. The zircon is approximately 600 microns long. The picture was taken with cross-polarized light at a magnification of 40x.

inset fig. 38). Smaller black amphibole-rich dikes, on the order of a centimeter or two in width and with more variable orientation can be observed cross-cutting the host gneisses on the rocky hillside just above the trenches. These likely represent a distinct generation of dikes whose size and mode of emplacement may indicate a minor and later secondary event (e.g. Tyler, 1979).

Age of Primary Mineralization

The mineralization history at the Powers farm is complex and cannot yet be fully described and documented. Many of the collecting sites display alteration of primary minerals into secondary ones including the formation of secondary quartz and calcite, widespread uralitization of large pyroxenes, and the introduction of minor sulfides and rare-earth-element minerals that are described elsewhere within this book. Here the focus is on the age and origin of the primary crystallization event responsible for the large, well-formed tourmaline crystals characteristic of the trenches and hilltop collecting sites. How this information applies to the remainder of the site and other types of mineralization remains to be investigated.

Field relationships suggest the dikes observed near the banks of Leonard Brook were emplaced after deformation had imparted a strong fabric to the host rocks (fig. 38)—an observation supported by U/Pb analyses of zircons (fig. 39) recovered from a sample of tourmaline-quartz-calcite-pyroxene dike rock that yielded an age of 1,158 (± 3) million years ago (Chiarenzelli et al., 2014a). This age falls within, or just after, the lower age limit of melting associated with the Shawinigan Orogeny in the Adirondack Lowlands (about 1,180 to 1,160 million years ago; Heumann et al., 2006). This also suggests that the tourmaline-bearing dikes at Leonard Brook were intruded at or near peak metamorphic conditions, but after most or all of the deformation in the host rocks was imparted. This further explains why the dikes at Leonard Brook are so straight, sharply bounded, and nearly parallel to one another.

Origin of the Powers Farm Black Tourmaline Locality

The geochronology presented above has established that at the time of the intrusion of the tourmaline-bearing dikes along the banks of Leonard Brook, the entire area was at or near peak metamorphic conditions—conditions that were hot enough to cause widespread ductile deformation, recrystallization and metamorphic mineral growth in local rocks, and partial melting in rocks of appropriate composition. The metamorphic events were, at least in part, driven by the large amount of igneous rock intruded into the region at this time, ranging in age between 1,200 and 1,150 million years ago (McLelland, et al., 1988; Peck et al., 2013). The amount of igneous activity in the Lowlands was less than in the Highlands (fig. 34); however, it would have served as a major source of heat.

If at least some of the tourmaline-bearing rocks at Powers farm are igneous in origin, then the question becomes: How did the melt from which they crystallized originate? The relatively unusual boron-rich composition of the Powers farm mineralized dikes is intriguing. Questions as to the source of the boron, physicochemical nature of the melt, and mode of emplacement remain to be answered. Some clues and constraints come from the surrounding area. Tourmaline is an abundant mineral in the Adirondack Lowlands, but far less so in the Highlands. A major belt of tourmaline-rich gneisses occurs within the Lower Marble, extending for more than 50 kilometers northwest of Route 11 (Brown and Ayuso, 1985). These rocks contain upwards of 50 percent or more tourmaline and their origin is also unclear, although some recent progress has been made in their interpretation (Chiarenzelli et al., 2014b). What is clear is that metasedimentary rocks in the surrounding area contain a large amount of boron essential to the formation of tourmaline.

Regional Considerations of Powers Farm Study
As previously mentioned, a spatial association between numerous pegmatitic calc-silicate mineral localities and the trace of the C-CSZ has been documented. More specifically, the vast majority of these localities are developed within, or adjacent to, marbles and rusty weathering, diopside-bearing, calc-silicate gneisses. Many, but not all, of these localities also contain large segregations of tourmaline in addition to assemblages that generally include diopside and/or tremolite, calcite, quartz, phlogopite, microcline, scapolite, sulfides, and less common minerals such as danburite. Another key to the formation of these deposits was elevated temperatures imparted by mountain-building processes and igneous intrusion. Given the boron-rich composition of the local rocks, the area was fertile for tourmaline production.

Summary
The Powers farm tourmaline locality is one of many such mineral occurrences along the C-CSZ, the boundary between the Adirondack Highlands and Lowlands. Unique in its assemblage and dominated by large, well-formed, black tourmaline crystals, it shares many similarities with other pegmatitic calc-silicate mineral locations in the Adirondack Region and in parts of the Central Metasedimentary Belt of the Grenville Province in Canada. It is hosted in a highly deformed and variable assemblage of metasedimentary rocks that are part of the widespread Grenville Supergroup that occurs as part of the basement rocks in southern Ontario, Quebec, and New York. Along Leonard Brook, four parallel dikes of tourmaline-bearing pegmatites cross-cutting the foliation in host gneisses have been dated and confirm an intrusive origin (1,158 ± 3 million years; Chiarenzelli et al., 2014a). The timing indicates emplacement at the end of the Shawinigan Orogeny and is coincident with the intrusion of a vast amount of igneous rocks throughout the Adirondack Highlands and Lowlands (McLelland et al., 1988; McLelland et al., 2013; Peck et al., 2013).

LITERATURE CITED

Brown, C. E. and R. A. Ayuso. "Significance of tourmaline-rich rocks in the Grenville Complex of St. Lawrence County, New York." *United States Geological Survey Bulletin* (1985): 1626-C.

Brown, Grenville. "Stratigraphy and Structure in the Balmat-Edwards District, Northwest Adirondacks, New York." *Geological Society of America Bulletin* 67 (1956): 1599–1622.

Chamberlain, S. C. "Excavation of a pegmatite dike on the Bower Powers farm, Pierrepont, St. Lawrence County, New York." *Rocks & Minerals* 82 (2007): 233–234.

Chamberlain, S. C. and G. W. Robinson. *The Collector's Guide to the Minerals of New York State*. Atglen, Pa.: Schiffer Publishing, 2013.

Chiarenzelli, J., M. Lupulescu, B. Selleck, G. Robinson, M. Pecha, H. Hagen-Peter, and T. Lockwood. "Geochronology of classic mineral deposits of St. Lawrence County, New York." *Geological Society of America Abstracts with Programs* 46 (2014a): 752.

Chiarenzelli, J., D. Kratzmann, B. Selleck, and W. deLorraine. "Age and provenance of Grenville Supergroup rocks, Trans-Adirondack Basin, constrained by detrital zircons." *Geology* 42 (2014b):183–186.

Geraghty, E. P., Y. W. Isachsen, and S. F. Wright. "Extent and character of the Carthage-Colton mylonite zone, northwest Adirondacks, New York." US Nuclear Regulatory Commission, NUREG/CR-1865 (1981).

Heumann, M. J., M. E. Bickford, B. M. Hill, J. M. McLelland, B. W. Selleck, and M. J. Jercinovic. "Timing of anatexis in metapelites from the Adirondack lowlands and southern highlands; a manifestation of the Shawinigan Orogeny and subsequent anorthosite-mangerite-charnockite-granite magmatism." *Geological Society of America Bulletin* 118 (2006): 1283–1298.

Isachsen, Y. W. and D. W. Fisher. "Geologic map of New York: Adirondack shee." New York State Museum, Map and Chart Series 15, scale 1:250,000 (1970).

Logan, W. E., Geology of Canada, Geol. Survey of Canada, Report of Progress from Commencement to 1863 (1863).

Lucas, S. B., and M. R. St-Onge (eds.), "Geology of the Precambrian Superior and Grenville Provinces and Precambrian Fossils in North America," Geological Survey of Canada, Ottawa (1998).

McLelland, J., B. Selleck, and M. E. Bickford, "Tectonic "Evolution of the Adirondack Mountains and Grenville Orogen Inliers within the USA," *Geoscience Canada* 40 (2013): 1–34.

McLelland, J., J. Chiarenzelli, P. Whitney, and Y. Isachsen. "U-Pb zircon geochronology of the Adirondack Mountains and implications for their geologic evolution." *Geology* 16 (1988): 920–924.

Peck, W., B. Selleck, M. Wong, J. Chiarenzelli, K. Harpp, K. Hollocher, J. Lackey, J. Catalano, S. Regan, and A. Stocker. "Orogenic to postorogenic (1.20–1.15 Ga) magmatism in the Adirondack Lowlands and Frontenac terrane, southern Grenville Province, USA and Canada." *Geosphere* 9 (2013):1637–1663.

Rivers, T. "Assembly and preservation of lower, mid, and upper orogenic crust in the Grenville Province—Implications for the evolution of large hot long-duration orogens." *Precambrian Research* 167 (2008): 237–259.

Roden-Tice, M. K. and S. J. Tice. "Regional Scale Mid-Jurassic to Late Cretaceous Unroofing from the Adirondack Mountains through Central New England based on Apatite Fission-Track and (U-Th)/He Thermochronology." *Journal of Geology* 113 (2005): 535–552.

Selleck, B., J. M. McLelland, and M. E. Bickford. "Granite emplacement during tectonic exhumation: The Adirondack example." Geology 33 (2005):781–784.

Streepey, M. M., E. L. Johnson, K. Mezger, and B. A. van der Pluijm. "Early history of the Carthage-Colton shear zone, Grenville Province, northwest Adirondacks, New York (USA)." *Journal of Geology* 109 (2001): 479–492.

Tyler, R. D. "Chloride metasomatism in the southern part of the Pierrepont Quadrangle, Adirondack Mountains, New York [Ph.D. thesis]." Binghamton, NY: State University of New York at Binghamton, 1979.

CHAPTER 3

THE SITES AND THEIR MINERALS

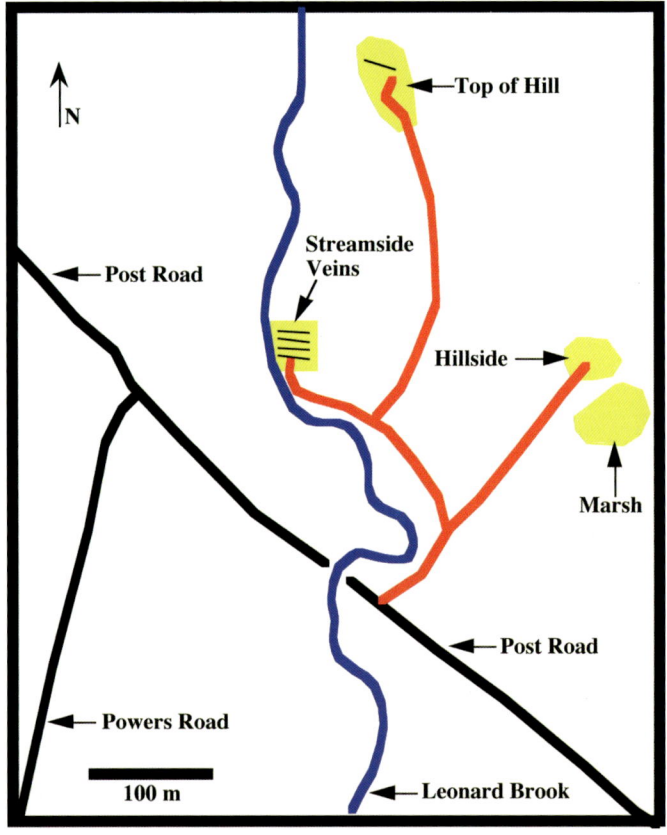

Figure 40. Site map. The principal mineralized areas are shown in yellow. The four streamside veins, shown by black lines, are, from south to north: Phosphate Vein, Middle Vein, Waddell Vein, and Wallace-Carlin Vein. The Smoky Quartz Vein is the black line in the middle of the Top of Hill site. The Marsh is also sometimes called the Swamp.

The black tourmaline locality covers many acres east of Leonard Brook and north of Post Road in the town of Pierrepont, St. Lawrence County, New York. If all the vegetation and topsoil were removed, it might reveal that the occurrence is a single large mineralized area. Historically, however, the various sites with mineralization have been discovered separately, accessed separately, and generally regarded as different sites on the greater locality. Therefore, they are treated individually here.

Most of the minerals at this occurrence formed as part of a pegmatite emplaced in Precambrian time. Some of these have been altered by retrograde metamorphism and weathering. However, a second period of hydrothermal activity occurred in open fractures forming additional minerals. In the following descriptions, these are described as late-stage minerals.

The mineral identifications given in this chapter are as accurate as can be determined now. Some have been identified by sight (e.g. quartz), many have been identified from physical characteristics and semi-quantitative chemical analysis with a semi-quantitative scanning electron microscope (SEM/EDS), some have been identified from physical characteristics and X-ray diffraction (XRD), and a few have been analyzed with both SEM/EDS and XRD. Late-stage, rare-earth minerals were found by SEM back-scattered electron imagery and identified by their morphology and chemical composition. The black tourmaline has been analyzed in detail by a variety of techniques (see chapter 5).

Top of the Hill

The exposures on top of the hill above Leonard Brook constitute the original locality discovered in 1859. This area has produced vast numbers of specimens and is still very productive more than 150 years after its discovery. This site is reached most easily by following the path leading to the left out of the entrance pasture on Post Road. The next path to the left goes to the streamside trenches. Staying to the right traverses an open woodland along an old farm road. The second path to the left leads to the top of the hill locality. Proceeding along this path, diggings in the woods are obvious on both sides of the path. At the top of a gentle rise, there are a series of small, interconnected clearings. The first is to the left of the path, the second is to the right at the end of the path, and the third, adjacent to the second, is just over the top of the hill. An open vein of mineralization (Smoky Quartz Vein) was discovered about forty years ago and separates the first clearing from the second. It will be discussed separately below. These clearings have been collected extensively, so the detailed topography changes from year to year.

Crystallized minerals of interest to collectors occur on top of the hill at contacts between calcite and metasedimentary rocks of variable appearance (mostly made up of tourmaline, quartz, and mica) or completely frozen in coarse-grained calcite. As the enclosing calcite has been dissolved away in the groundwater, crystals and clusters of crystals have ended up loose in the soil. The complex geometry of the calcite-filled veins and large calcite masses has been further obscured by the extensive collecting activities and backfilling that have occurred since the locality was discovered.

Figure 41. Rusty area on Top of Hill. June 12, 2014. SCC

Minerals
The following descriptions include all the verified species from the top of the hill, excluding those found in the Smoky Quartz Vein.

Actinolite, $Ca_2Mg_{<4.5}Fe^{2+}_{>0.5})Si_8O_{22}(OH)_2$ to $Ca_2Mg_{2.5}Fe^{2+}_{2.5})Si_8O_{22}(OH)_2$ occurs as black bladed prismatic crystals to 12 cm. The freestanding crystals from the rusty area on top of the hill contain just enough iron to qualify as actinolite.

Figure 42. Actinolite. 7 cm. Collected by George Robinson in 1985; now at the Canadian Museum of Nature, #51759. *GWR*

Calcite, $CaCO_3$, occurs as massive sheets and vein-fillings of coarse-grained creamy white to tan rock. Radiaxial calcite occasionally coats the surfaces of wall minerals in weathered veins. Calcite crystals found on top of the hill are part of the fracture-filling mineralization described below.

Figure 43. Calcite, quartz. 5.4 cm. Collected by Steve Chamberlain in 1982 and in his collection, #5458. *SCC*

Chalcopyrite, $CuFeS_2$, occurs rarely as sharp crystals to 2.7 cm encased in calcite. Usually these have a dull tan rind with small spots of green malachite.

Figure 44. Chalcopyrite, malachite. 2.7 cm. Collected by Mike Walter in 2013 and in his collection, #00025. *MW*

Chlorite group minerals occur as alteration products of other species, typically phlogopite, diopside, and marialite. We have not attempted to determine which species occur in these alteration mixtures, except that clinochlore is often an alteration product of marialite.

Diopside, $CaMgSi_2O_6$, occurs as blocky crystals to 8 cm made up of the three pinacoids, often completely or partially altered to mixtures of tremolite and talc, i.e. uralite. Less frequently diopside forms very dark green glassy prismatic crystals to 3 cm or more consisting of lateral pinacoids and several monoclinic prisms, giving them more complex terminations.

Figure 45. Diopside. 3.5 cm. Collected by Steve Chamberlain in 1981 and in his collection, #17361. SCC

Figure 46. Diopside. 5.8 cm. Collected by Mike Walter in 2013 and in his collection, #00874. MW

Fluorapatite, $Ca_5(PO_4)_3F$, occurs as tan, pale green, gray, and brown prismatic crystals consisting of the hexagonal prism and basal pinacoid. Frequently the edges of the terminations are beveled by small hexagonal pyramid faces. Fluorapatite occurs frozen in calcite and with other species lining the calcite/rock contacts. Most crystals are smaller than 1 cm but are sometimes up to 5 cm.

Figure 47. Fluorapatite. 3.6 cm. Collected by Donald Carlin Jr. in 2009, now in the Chamberlain collection, #22062. SCC

Figure 48. Fluorapatite, tourmaline. 1.5-cm crystal. Collected by Jay Walter in 1995 and in his collection. MW

Figure 49. Fluorapatite, quartz. 1.7-cm crystals. Collected by Donald Carlin Jr. in 2012, now in the Chamberlain collection, #27537. SCC

Goethite, FeO(OH), at this site is probably always the result of oxidative weathering of iron sulfides, principally pyrite. It occurs as sharp euhedral pseudomorphs after pyrite and as ocherous brown coatings and masses.

Figure 50. Goethite, marcasite, quartz. 5.4 cm. Collected by Donald Carlin Jr. in 2009, now in the Chamberlain collection, #24125. SCC

Magnetite, Fe_3O_4, occurs as black lustrous metallic octahedral crystals up to more than 2 cm, although most crystals are smaller.

Figure 51. Magnetite. 2.2 cm. Collected by George Robinson in 1978, now in the Chamberlain collection, #1846. SCC

Malachite, $Cu_2(CO_3)(OH)_2$, is an uncommon green coating resulting from the weathering of chalcopyrite.

Marialite, $Na_4Al_3Si_9O_{24}Cl$, occurs as tan or gray prisms to several cm, often associated with phlogopite and, more rarely, titanite. The composition of this scapolite is intermediate between marialite and meionite, but is on the sodium-rich end of the series. Crystals typically have two tetragonal prisms, one or more tetragonal bipyramids, and basal pinacoids.

Figure 52. Marialite. 5.2 cm. Collected by Ivan McIntosh before 1978, now in the Chamberlain collection, #1723. *SCC*

Figure 53. Marialite, phlogopite. Collected by Steve Chamberlain in 1976 and in his collection, #22314. *SCC*

Phlogopite, $KMg_3AlSi_3O_{10}(OH)_2$, occurs as dark brown crystals with prominent cleavage. Some form prismatic crystals that gradually taper to a termination and may be up to 15 cm. Others are equant or tabular. Most specimens show a cleavage face rather than a termination. Rarely, triplet twins have been found. This is an iron-rich phlogopite with traces of titanium and fluorine.

Figure 54. Phlogopite, calcite. 8.5 cm. Collected by Donald Carlin Jr. in 2011, now in the Chamberlain collection, #24631. SCC

Figure 55. Phlogopite. 14.7 cm. Collected by Mike Walter in 2014 and in his collection, #00870. MW

Figure 56. Phlogopite triplet twin. 8.5 cm. Collected by Mike Walter in 2013 and in his collection, #00682. *MW*

Figure 57. Phlogopite and calcite in "graphic intergrowth." 5 cm. Collected by Steve Chamberlain in 1975 and in his collection, #5047. *SCC*

Pyrite, FeS_2, occurs as anhedral and subhedral masses to several cm and as sharp cubic and octahedral crystals to 5 cm. Most samples are coated with goethite to some degree.

Pyrrhotite, $Fe_{1-x}S$, is a rare mineral on top of the hill. Sharp, prismatic metallic gold hexagonal crystals with bipyramidal terminations to 1.5 cm have been found associated with quartz. These are actually pyrite pseudomorphs after pyrrhotite.

Quartz, SiO_2, occurs as colorless to white to smoky brown crystals to 12 cm. Most crystals have tapered prism faces (Tessin habit) with striations and are translucent rather than transparent. More rarely, transparent crystals with flat prism faces have been found.

Figure 58. Quartz. 4 cm. Collected by Donald Carlin Jr. in 2012, now in the Chamberlain collection, #26074. *SCC*

Ready to Write a Book?

Our authors are as passionate as we are about providing new and intriguing perspectives on a variety of topics, both niche and general. If you have a fresh idea, we would love to hear from you, as we are continually seeking new authors and their work. Visit our website to view our complete list of titles and our current catalogs. Please visit our Author Resource Center on our website for submission guidelines, and contact us at proposals@schifferbooks.com or write to the address below, to the attention of Acquisitions.

Schiffer Publishing Ltd.

A family-owned, independent publisher since 1974, Schiffer has published thousands of titles on the diverse subjects that fuel our readers' passions. Explore our list of more than 5,000 titles in the following categories:

ART, DESIGN & ANTIQUES

Fine Art | Fashion | Architecture | Interior Design | Landscape | Decorative Arts | Pop Culture | Collectibles | Art History | Graffiti & Street Art | Photography | Pinup | Sculpture | Body Art & Tattoo | Antique Clocks | Watches | Graphic Design | Contemporary Craft | Illustration | Folk Art | Jewelry | Fabric Reference

MILITARY

Aviation | Naval | Ground Forces | American Civil War | Militaria | Modeling & Collectible Figures | Pinup | Transportation | World War I & II | Uniforms & Clothing | Biographies & Memoirs | Unit Histories | Emblems & Patches | Weapons & Artillery

CRAFT

Arts & Crafts | Fiber Arts & Wearables | Woodworking | Quilts | Gourding | Craft Techniques | Leathercraft | Carving | Boat Building | Knife Making | Printmaking | Weaving | How-to Projects | Tools | Calligraphy

TRADE

Lifestyle | Natural Sciences | History | Children's | Regional | Cookbooks | Entertaining | Guide Books | Wildlife | Tourism | Pets | Puzzles & Games | Movies | Business & Legal | Paranormal | UFOs | Cryptozoology | Vampires | Ghosts

MIND BODY SPIRIT

Divination | Meditation | Astrology | Numerology & Palmistry | Psychic Skills | Channeled Material | Metaphysics | Spirituality | Health & Lifestyle | Tarot & Oracles | Crystals | Wicca | Paganism | Self Improvement

MARITIME

Professional Maritime Instruction | Seamanship | Navigation | First Aid/Emergency | Maritime History | The Chesapeake | Antiques & Collectibles | Children's | Crafts | Natural Sciences | Hunting & Fishing | Cooking | Shipping | Sailing | Travel | Navigation

SCHIFFER PUBLISHING, LTD.
4880 Lower Valley Road | Atglen, PA 19310
Phone: 610-593-1777
E-mail: Info@schifferbooks.com
Printed in China

www.schifferbooks.com

Figure 59. Quartz. 8 cm. Collected by Mike Walter in 2014, now in the Chamberlain collection, #31535. *SCC*

Figure 60. Quartz, tourmaline, phlogopite. 5.8 cm. Collected by Mike and Jay Walter in 2013, now in the Chamberlain collection, #30805. *SCC*

Figure 61. Quartz, phlogopite. 7.2 cm. Collected by Donald Carlin Jr. in 2013, now in the Chamberlain collection, #22061. *SCC*

Figure 62. Quartz, phlogopite. 6.6cm. Collected by Steve Chamberlain in 1976 and in his collection, #1150. *SCC*

Figure 63. Quartz. 5.7 cm. Collected by Mike Walter in 2013 and in his collection, #00679. *MW*

Serpentine subgroup minerals, especially antigorite, occur as part of alteration mixtures in pseudomorphs after diopside and scapolite.

Talc, $Mg_3(Si_4O_{10})(OH)_2$, occurs as white to tan pseudomorphs after quartz, diopside, and marialite, as partial coatings on diopside, and as lattice works filling cleavage planes in calcite. Talc also occurs as gray to silver primary crystals to 1 cm. These wafer-like crystals can be found in association with uralite and are easy to mistake for phlogopite crystals.

Titanite, $CaTiOSiO_4$, occurs as brown to black flattened crystals to 1 cm, usually associated with uralite or marialite.

Figure 64. Titanite, uralite. 3.2 cm. Collected by Steve Chamberlain in 1986 and in his collection, #17042. *SCC*

Figure 65. Titanite, marialite. 7.2 cm. Collected by Margaret Norris in 1962, now in Chamberlain collection, #15888. *SCC*

Tourmaline occurs as splendent black crystals to 13 cm. Crystals are typically equant, without striations, and consist of trigonal and ditrigonal prisms terminated by relatively flat trigonal pyramids, occasionally with a basal pedion. Tabular crystals are relatively common; elongated, prismatic crystals, less so. Doubly terminated crystals often show hemimorphic development with steeper trigonal pyramids on one termination and flatter trigonal pyramids on the other. Chemical analysis shows that crystals are zoned with a layer of dravite, $Na(Mg_3)Al_6(Si_6O_{18})(BO_3)_3(OH)_3(OH)$, on the surface and a core of fluor-uvite, $Ca(Mg_3)MgAl_5(Si_6O_{18})(BO_3)_3(OH)_3F$. Details are discussed in chapter 5.

Figure 66. Tourmaline, goethite. 8.5 cm. Collected by Mike Walter, July 2004, now in Chamberlain collection, #15390. SCC

Figure 67. Tourmaline, quartz. 12-cm fov. Collected by George Robinson, c. 1972. Canadian Museum of Nature collection. *Jeff Scovil photo*

Figure 68. Tourmaline. 10 cm. Collected by Mike Walter in 2013 and in his collection, #00696. *MW*

Figure 69. Tourmaline. 5 cm. A. E. Seaman Mineral Museum collection. *GWR*

Figure 70. Tourmaline. 3.5 cm. Collected by Mike Walter in 2014 and in his collection, #00861. MW

Figure 71. Tourmaline. 12 cm. Collected by Mike Walter in 2013 and in his collection, #00792. MW

Figure 72. Tourmaline, quartz. 10 cm. Collected by George Robinson in the 1980s, now in the Chamberlain collection, #8602. SCC

Figure 73. Tourmaline, fluorapatite, phlogopite, quartz. 10 cm. Collected by Donald Carlin Jr. in 2011, now in the Chamberlain collection, #23903. *SCC*

Figure 74. Tourmaline, quartz. 8.5 cm. Collected by Vern Phillips in 1985, now in Chamberlain collection, #8097. *SCC*

Figure 75. Tourmaline, quartz. 4.4 cm. Collected by Vern Phillips in 1985, now in the Chamberlain collection, #8095. *SCC*

Figure 76. Tourmaline, quartz. 5.7 cm. Collected by Vern Phillips in 1985, now in the Chamberlain collection, #8096. *SCC*

Tremolite, $Ca_2Mg_5Si_8O_{22}(OH)_2$ to $Ca_2Mg_{4.5}Fe^{2+}_{0.5}Si_8O_{22}(OH)_2$, occurs most commonly as an epitactic overgrowth on diopside or altered diopside (uralite) of dark green elongated or equant prisms. Altered diopside crystals (uralite) typically contain tremolite in the mixture of minerals to which they have altered. More rarely, tremolite occurs as clusters of equant dark green crystals to 0.5 cm.

Vermiculite, $Mg_{0.7}(Mg,Fe^{2+},Al)_6(Si,Al)_8O_{20}(OH)_4 \cdot 8H_2O$, occurs as an alteration product of phlogopite. Generally, phlogopite is dark brown, elastic, and transparent in thin cleavages. When altered to vermiculite, sheets are brown, translucent to pearly, and inelastic. Phlogopite altered to clinochlore tends to be silvery to green and translucent to pearly and inelastic.

Pseudomorphs

The following descriptions include all the pseudomorphs from the top of the hill that have been recognized as such.

Chlorite ps. after phlogopite are relatively common. A cleavage sheet of many crystals of phlogopite shows a rim altered to chlorite, mostly clinochlore, which is gray to green, somewhat pearly, and inelastic. In some cases, the phlogopite has been completely altered.

Chlorite ps. after marialite occur in the vicinity of the Smoky Quartz Vein. They are typically greenish-gray, with typical scapolite crystal faces, but are extremely fragile. XRD reveals that these are principally clinochlore with lesser amounts of antigorite.

Figure 77. Chlorite after marialite. 6 cm. Collected by Mike Walter in 2012 and in his collection, #00085. *MW*

Goethite ps. after pyrite are typically altered to varying degrees from a thin brown surface layer with metallic pyrite still visible, to a complete replacement of the original pyrite.

Figure 78. Goethite after pyrite, 2.7 cm. Collected by Steve Chamberlain in 1983 and in his collection, #13677. *SCC*

Pyrite ps. after pyrrhotite are rare. They are spindle-shaped, tapered metallic golden crystals consisting of a hexagonal prism and steep hexagonal bipyramid. The surface texture is granular from the replacement of pyrrhotite by pyrite.

Quartz ps. after diopside are shaped like blocky diopside crystals, but have a white to gray coarsely granular replacement texture. Sometimes diopside crystals are partly replaced by greenish tremolite and talc and partly by white quartz.

Talc ps. after diopside are typically white or gray and have the blocky shape of pinacoidal diopside crystals or the prismatic shape of diopside crystals with more complex terminations.

Talc ps. after marialite are relatively rare. They are dark green, translucent crystals with the tetragonal prisms and bipyramids typical of scapolite.

Figure 79. Talc after marialite. 4 cm. Collected by Don Briggs in 1975, now in the Chamberlain collection, #1880. *SCC*

Talc ps. after quartz most frequently have the tapered striated prisms typical of Tessin habit quartz. The gray to white talc may be a thin coating on the surface, or a partial or complete replacement of the quartz.

Figure 80. Talc after quartz, Tourmaline. 7.5 cm. Collected by Mike Walter, Lucas Snell, and Gary Snell in 2014, now in the Mike Walter collection, #00853. *MW*

Figure 81. Talc after quartz. 2.5 cm. Collected by John Burrow in 1975, now in the Chamberlain collection, #31539. *SCC*

Tremolite ps. after diopside (uralite) usually are greenish gray and have the blocky pinacoidal shape of the original diopside crystals. Sometimes these are overgrown by arrays of glassy green tremolite crystals that were epitactic to the unaltered diopside. Many of them are mixtures of minerals including talc, quartz, and mica as well as tremolite.

Figure 82. Uralite, tourmaline. 18-cm fov. Collected by George Robinson in 1970, now in the Canadian Museum of Nature collection, #45422. GWR

Figure 83. Uralite, tourmaline. 8.5 cm. Collected by George Robinson in 1966, now in the Chamberlain collection, #6350. SCC

Chapter 3: The Sites and Their Minerals

Figure 84. Uralite, tourmaline. 12 cm. Collected by Mike Walter in 2013 and in his collection, #00701. *MW*

Figure 85. Uralite. 5.4 cm. Collected by Mike Walter in 2013 and in his collection, #00648. *MW*

55

Figure 86. Uralite. 5.6 cm. Collected by Steve Chamberlain in 1981 and in his collection, #31554. *MW*

Figure 87. Uralite, talc. 6.2 cm. Collected by Mike Walter on August 26, 2013, now in the Chamberlain collection, #30812. *SCC*

Vermiculite ps. after phlogopite resemble chlorite ps. after phlogopite except that cleavage sheets are brown, often pearly, and inelastic.

Smoky Quartz Vein

In the early 1970s, a vein with fracture-filling mineralization was discovered on top of the hill where the first clearing to the left of the path merges into the second clearing at the path's end. This vein, subsequently named the Smoky Quartz Vein, was a fracture filling without a calcite core. Its initial excavation formed a linear trench that opened on the steep hillside falling into the creek on one end and running almost to the entry path on the other. In the first few years, it reached a depth of 6 to 8 feet below the surface. Subsequent collecting in the dumps quickly obscured the original trench so that today, nothing resembling the original outlines of the vein is visible.

This vein appears to be the result of late-stage hydrothermal mineralization of an older fracture in the Precambrian pegmatite. The principal minerals lining the open vein were transparent pale smoky quartz grading into darker, almost black quartz and ivory to tan calcite crystals.

Collecting activity uphill (northeast) from the Smoky Quartz Vein has occasionally revealed small areas of similar open-space mineralization with quartz and calcite crystals and occasional rare-earth minerals. This suggests that the open fracture system preceding the mineralization of the Smoky Quartz Vein may have had some smaller lateral branches that were also mineralized by hydrothermal solutions.

Figure 88. Approximate location of Smoky Quartz Vein. June 12, 2014. *SCC*

Minerals

The following list and descriptions include all the verified species of late-stage minerals found in the Smoky Quartz Vein and its lateral branches that are part of the hydrothermal mineralization.

Allanite-(Ce), $Ca(Ce,REE)Al_2Fe^{2+}[Si_2O_7][SiO_4]O(OH)$, occurs as black prismatic crystals smaller than 1 mm on calcite crystals from lateral branches.

Figure 89. Allanite-(Ce), Calcite, quartz. Lateral vein. 4.6 cm. Collected by Mike Walter in 2002 and in his collection, #00037. *MW*

Figure 90. Allanite-(Ce), calcite. 8.6 cm. Lateral vein. Collected by Mike Walter in 2000 and in his collection, #00036. *MW*

Barite, $BaSO_4$, sparsely occurs as brown bladed crystals with feathered terminations to 3 cm, and also as minute tabular crystals embedded in calcite crystals.

Figure 91. Barite. 2.9 cm. Collected by Donald Carlin Jr. in 2009, now in the Chamberlain collection, #24369. *SCC*

Bornite, Cu_5FeS_4, occurs as minute crystals observable with an SEM in late-stage calcite.

Calcite, $CaCO_3$, is abundant as white to tan rhombohedral crystals to 10 cm. Some form twins similar to those from the classic Rossie lead mines. Minute calcite crystals of both rhombohedral and scalenohedral habit occur interspersed in the fine-grained gray chamosite that also contains rare-earth-element minerals.

Figure 92. Calcite, magnetite. Lateral vein. 3.8 cm. Collected by Mike Walter in 2003 and in his collection, #00038. *MW*

Figure 93. Calcite twin. 3 cm. Collected by Donald Carlin Jr. in 2009, now in the Chamberlain collection, #24371. *SCC*

Chalcocite, Cu_2S, occurs as minute crystals observable with an SEM in late-stage calcite.

Chalcopyrite, $CuFeS_2$, occurs as sharp crystals to 1 cm, often partially coated by green malachite.

Figure 94. Chalcopyrite, calcite, quartz. 7 cm. Collected by Donald Carlin Jr. on September 25, 2009, now in the Chamberlain collection, #22400. *SCC*

Chamosite, $(Fe,Al,Mg)_6(Si,Al)_4O_{10}(OH)_8$, forms gray masses at the bases of calcite and quartz crystals. When viewed with the SEM, the chamosite shows rosettes of platy crystals. Calcite, monazite-(Ce), synchysite-(Ce), and xenotime-(Y) occur embedded in the massive chamosite.

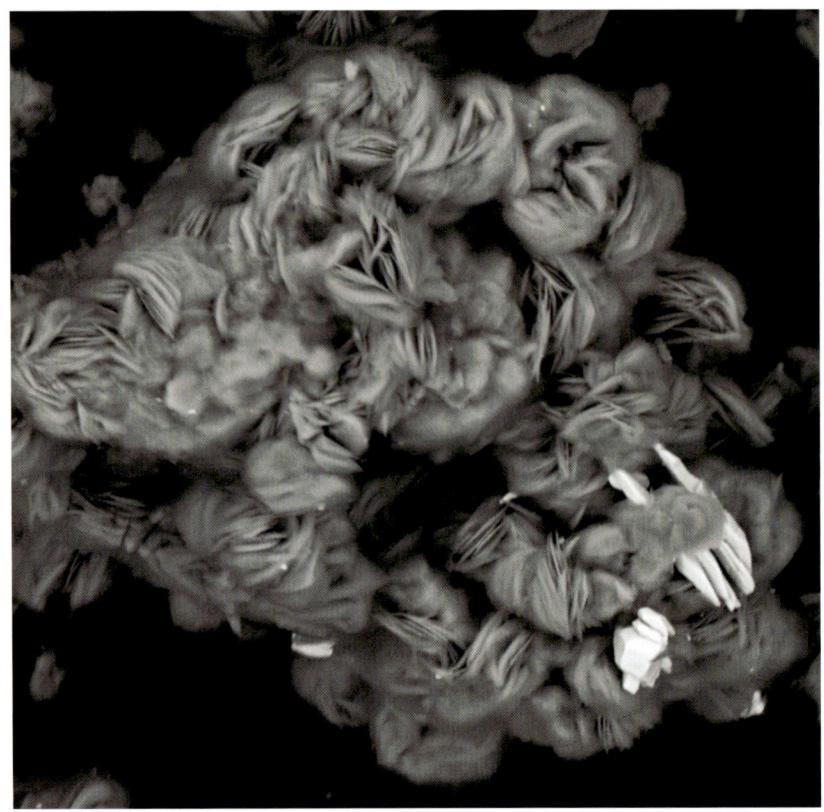

Figure 95. Chamosite, monazite-(Ce). 171-μm fov. Chamberlain collection. *SCC & DGB*

Magnetite, Fe_3O_4, occurs as black octahedral crystals on calcite crystals in lateral branches and as minute crystals observable with an SEM in late-stage calcite.

Malachite, $Cu_2(CO_3)(OH)_2$, is a common green coating, often botryoidal, on chalcopyrite crystals and adjacent calcite crystals.

Figure 96. Malachite, chalcopyrite, quartz, calcite. 6.2 cm. Collected by Donald Carlin Jr. in 2009, now in the Chamberlain collection, #22504. *SCC*

Figure 97. Malachite, chalcopyrite. 2.2-mm crystal. Chamberlain collection, #20070. *SCC*

Monazite-(Ce), $(Ce,La,Nd,Th)(PO_4)$, occurs as microscopic elongated prismatic crystals to 10 μm embedded in chamosite and in late-stage calcite.

Figure 98. Monazite-(Ce). 114-μm fov. Chamberlain collection. *SCC & DGB*

Quartz, SiO_2, is common as terminated prismatic crystals to 12 cm. These range from colorless to pale brown to almost black and from transparent to translucent. They are among the best quartz crystals ever found in northern New York.

Figure 99. Quartz. 11 cm. Collected by Nick Rochester in 1978, now in the Chamberlain collection, #7727. *SCC*

Figure 100. Quartz. 7.5-cm fov. Collected by Ivan McIntosh in 1978, now in the Chamberlain collection, #1701. *SCC*

Chapter 3: The Sites and Their Minerals

Figure 101. Quartz. 5.8 cm. Collected by Donald Carlin Jr. in 2009, now in the Chamberlain collection, #22407. SCC

Figure 102. Quartz. 10.6-cm crystal. Collected by George Robinson in 1978, now in the Chamberlain collection, #2278. SCC

Figure 103. Quartz, calcite, 2.5-cm crystal. Collected by Donald Carlin Jr. in 2009, now in the Chamberlain collection, #21005. SCC

63

Figure 104. Quartz, calcite. 13 cm. Collected by Nick Rochester in 1978, now in the Canadian Museum of Nature collection, #50510. GWR

Figure 105. Quartz, tourmaline. 7 cm. Chamberlain collection, #2764. SCC

Chapter 3: The Sites and Their Minerals

Figure 106. Quartz. 7.5 cm. Collected by M. D'Amore, now in the Chamberlain collection, #6322. *SCC*

Figure 107. Quartz. 7 cm. Collected by Vern Phillips in 1983, now in the Chamberlain collection, #9305. *SCC*

65

Figure 108. Quartz, calcite. 14.5 cm. Collected by Donald Carlin Jr. in 2010, now in the Chamberlain collection, #22967. SCC

Figure 109. Quartz, tourmaline. Lateral vein. 4 cm. Collected by Jay Walter in 1995 and in his collection. MW

Figure 110. Quartz, tourmaline. Lateral vein. 4.3 cm. Collected by Vern Phillips in 1984, now in the Chamberlain collection, #9345. *SCC*

Figure 111. Quartz, tourmaline. Lateral vein. 5.7 cm. Collected by Vern Phillips in 1984, now in the Chamberlain collection, #9306. *SCC*

Figure 112. Quartz, magnetite. 8-cm fov. Collected by Donald Carlin Jr. 2009, now in the Chamberlain collection, #21001. *SCC*

Synchysite-(Ce), Ca(Ce,Y,Nd,La)(CO$_3$)$_2$F, forms hexagonal prismatic crystals terminated by the basal pinacoid and may be tabular or elongated to 70 μm. Crystals occur embedded in chamosite and in late-stage calcite.

Figure 113. Synchysite-(Ce), calcite. 38-μm fov. Chamberlain collection. *SCC & DGB*

Synchysite-(Y), Ca(Y,Ce,Nd,La)(CO$_3$)$_2$F, occurs as melted-looking, chalice-shaped crystals to 20 μm embedded in chamosite.

Talc, Mg$_3$(Si$_4$O$_{10}$)(OH)$_2$, occurs as rough gray to silvery micaceous crystals to 1 cm.

Tourmaline rarely occurs as splendent black crystals on quartz and calcite crystals. We interpret these as having been loosened from the fracture walls during hydrothermal mineralization and having become embedded in the late-stage mineralization.

Figure 114. Tourmaline, quartz, calcite. 4.5 cm. New York State Museum collection, #21750. *Stephen Nightingale photo*

Figure 115. Tourmaline, quartz, calcite. 8 cm. Collected by Jay Walter in 1992 and in his collection. *MW*

Xenotime-(Y), $Y(PO_4)$, occurs as simple tetragonal prisms and bipyramids to 10 μm embedded in chamosite and in late-stage calcite.

Figure 116. Xenotime-(Y) & synchysite-(Ce). 22-μm crystal, Chamberlain collection. *SCC & DGB*

Streamside Veins

For more than 100 years, the pegmatite veins along the stream lay unnoticed. Sometime in the late 1950s, a vein of tourmaline and quartz running from the side of the hill into Leonard Brook was discovered and exploited through the 1960s, after which this part of the locality lay fallow until 2004, when extensive mining for specimens began along the Waddell Vein. Early in 2008, a second parallel vein, the Phosphate Vein, was discovered a short distance to the south. This vein was excavated as deep as the remaining calcite core, almost to Leonard Brook to the west, and to the east until it pinched out. Later in 2008, a third vein, the Middle Vein, was noticed midway between the Phosphate Vein and the Waddell Vein and excavated. Also late in 2008, a fourth vein, the Wallace-Carlin Vein, was discovered a short distance to the north of the Waddell Vein and was excavated through 2009.

The streamside veins can be reached by bearing to the left on the paths leaving the entrance pasture on Post Road. Turning left at the first branch in the path brings one directly to the four nearly parallel streamside veins. These run roughly perpendicular to Leonard Brook and are encountered as one follows the path in the order: Phosphate Vein, Middle Vein, Waddell Vein, and Wallace-Carlin Vein.

Phosphate Vein

The Phosphate Vein is a Precambrian quartz-tourmaline-pyroxene pegmatite vein emplaced in a vertical fracture in metasedimentary gneiss. Originally, the vein had a solid calcite core, but this has weathered to a depth of more than 2 meters. The Precambrian minerals exposed on the vein walls include tourmaline, Tessin-habit quartz, phlogopite, and fluorapatite. Subsequent alteration of these minerals formed pseudomorphs and occurred, at least in part, after the calcite core had dissolved away. Sometime later, hydrothermal solutions deposited quartz, microcline, chamosite, sphalerite, chalcopyrite, talc, and rare-earth-element minerals.

Figure 117. Phosphate Trench during excavation. July 15, 2008. *SCC*

Figure 118. Phosphate Trench. June 12, 2014. *SCC*

Minerals

Allanite-(Ce), $Ca(Ce,REE)Al_2Fe^{2+}[Si_2O_7][SiO_4]O(OH)$, occurs as late-stage, microscopic crystals associated with chamosite, quartz, and microcline, all of which are best observed with an SEM.

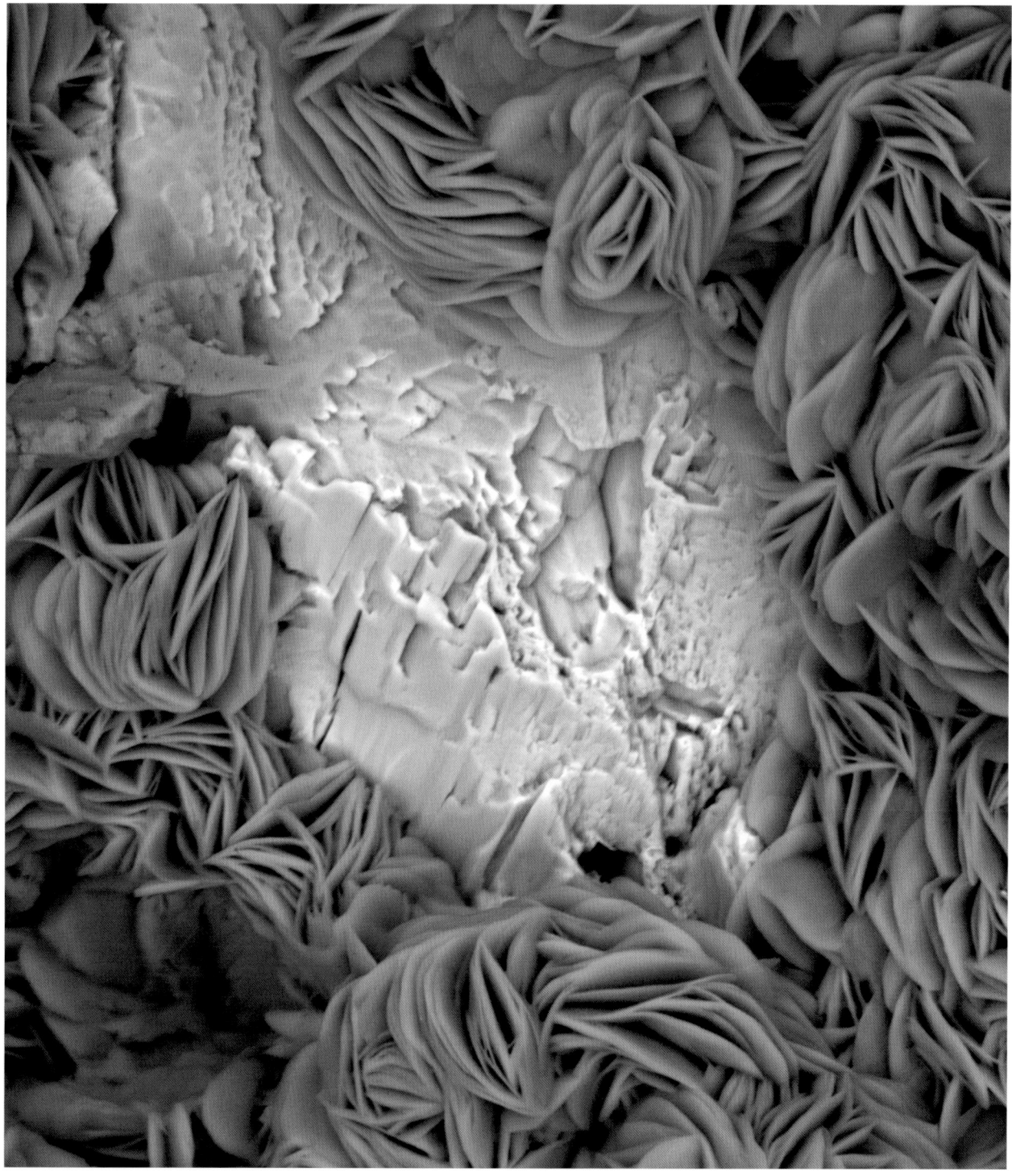

Figure 119. Allanite-(Ce), chamosite. 52-μm fov. Chamberlain collection. *SCC & DGB*

Calcite, $CaCO_3$, occurs as the tan core of the vein and also rarely as white to gray rhombohedral late-stage crystals on vein walls.

Chalcopyrite, $CuFeS_2$, occurs as late-stage iridescent crystals to 1 mm associated with late-stage talc crystals.

Chamosite, $(Fe,Al,Mg)_6(Si,Al)_4O_{10}(OH)_8$, forms late-stage, microscopic rosettes of flattened crystals observable with an SEM associated with quartz and microcline.

Chlorite, probably mostly clinochlore, occurs as a silvery or green alteration product of phlogopite and as one component of altered scapolite crystals.

Cordierite, $Mg_2Al_4Si_5O_{18}$, is found as late-stage minute fibrous crystals in fractures in tourmaline.

Diopside, $CaMgSi_2O_6$, occurs as prismatic green crystals to several centimeters. Many diopside crystals have been coated with or replaced by quartz.

Goethite, $FeO(OH)$, forms coatings on pyrite crystals and stains and earthy masses on many specimens.

Gold, Au, occurs as 2-μm sized grains observed in thin sections of Tessin-habit quartz.

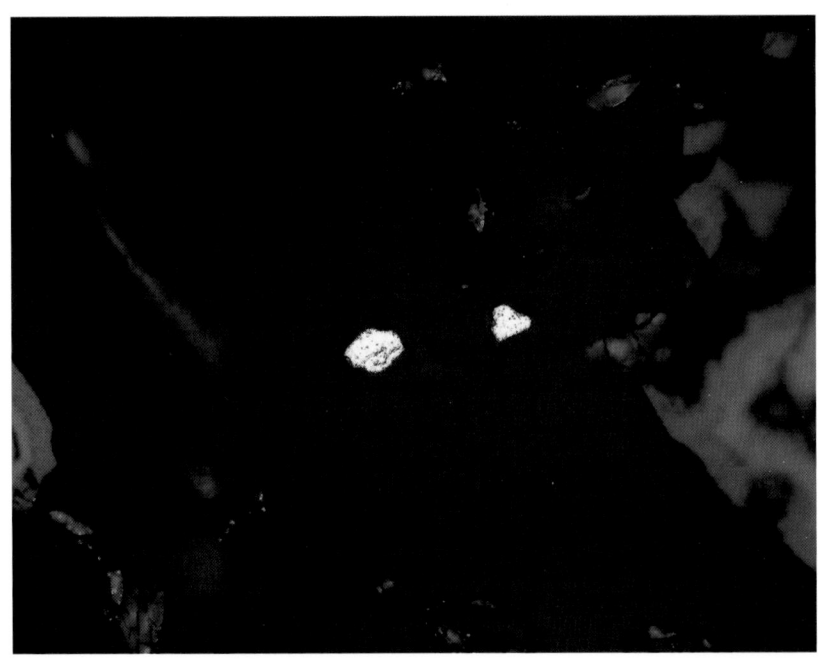

Figure 120. Gold, 2-μm grains. Collected by Marian Lupulescu in 2008. ML

Fluorapatite, $Ca_5(PO_4)_3F$, occurs commonly in prismatic crystals to 7 cm. Tan to brown hexagonal prisms are terminated by basal pinacoids and often beveled by hexagonal pyramids. The abundance of fluorapatite gave rise to the name Phosphate Vein.

Figure 121. Fluorapatite. 7.3 cm. Collected by Mike Walter in 2008 and in his collection, #00093. *MW*

Figure 122. Fluorapatite, phlogopite. 4.9 cm. Collected by Mike Walter in 2008, now in the Chamberlain collection, #19159. *SCC*

Figure 123. Fluorapatite (tabular). 3.2 cm. Collected by Mike Walter in 2008 and in his collection, #00055. *MW*

Figure 124. Fluorapatite, phlogopite. 7.4 cm. Collected by Mike Walter in 2008 and in his collection, #00053. *MW*

Marcasite, FeS_2, occurs sparsely, probably as a late-stage mineral, as silvery, flattened, orthorhombic prisms to several mm.

Figure 125. Marcasite. 2.6-mm fov. Chamberlain collection, #19195. *SCC*

Microcline, $KAlSi_3O_8$, occurs as late-stage, pseudorhombohedral, white to tan crystals to several mm, and as microscopic, etched crystals with chamosite and quartz, observable with an SEM.

Figure 126. Microcline. 4 mm-fov. Chamberlain collection. *SCC*

Figure 127. Microcline, chamosite. 0.48-mm fov. Chamberlain collection. *SCC & DGB*

Phlogopite, $KMg_3AlSi_3O_{10}(OH)_2$, forms pseudohexagonal, brown prisms to 6 cm. It is sometimes altered to clinochlore, vermiculite, or quartz.

Figure 128. Phlogopite. 5.5 cm. Collected by Steve Chamberlain in 2008 and in his collection, #18967. SCC

Figure 129. Phlogopite, quartz. 2.6 cm. Chamberlain collection, #19236. SCC

Pyrite, FeS$_2$, occurs in equant, or flattened, cubic crystals to several cm, often coated with goethite, and as complex, metallic, late-stage crystals to several mm.

Figure 130. Pyrite. 3.7-mm fov. Collected by Mike Walter in 2008, now in the Chamberlain collection. *SCC*

Quartz, SiO$_2$, occurs as white to gray crystals of Tessin habit to 16 cm, as late-stage, sharp, prismatic, transparent to white crystals, and as microscopic, late-stage, skeletal crystals observable with an SEM associated with chamosite and microcline. Quartz is also a pseudomorphous replacement of phlogopite and diopside.

Figure 131. Quartz, quartz after phlogopite. 5.5 cm. Collected by Mike Walter in 2008, now in the Chamberlain collection, #19190. *SCC*

Figure 132. Quartz, chamosite. 0.6-mm fov. Chamberlain collection. *SCC & DGB*

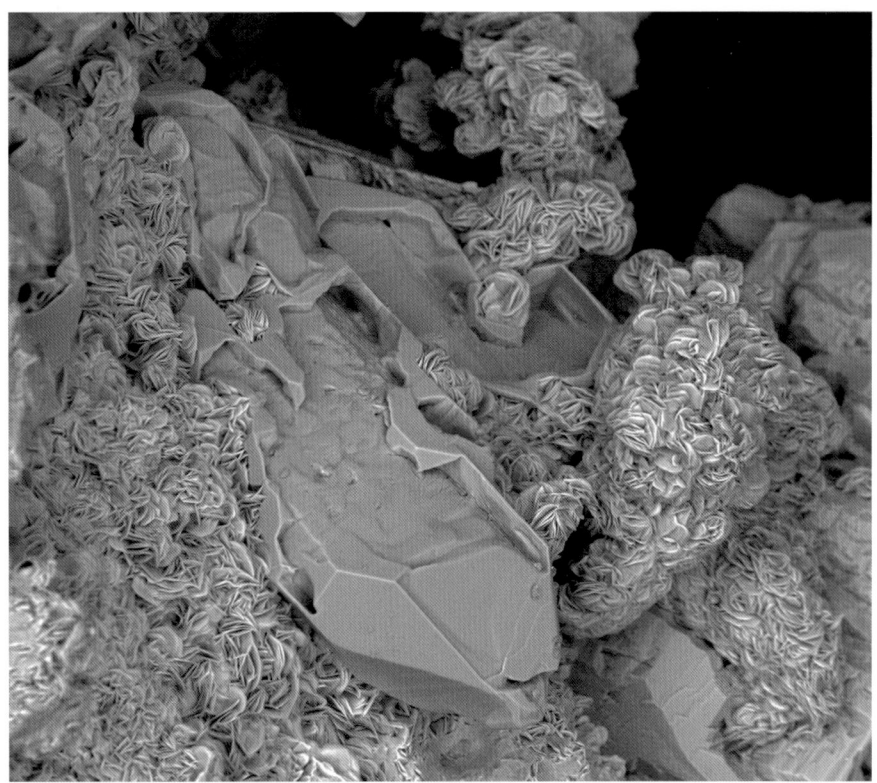

Figure 133. Quartz, synchysite-(Ce). 59 μm-fov. Chamberlain collection. *SCC & DGB*

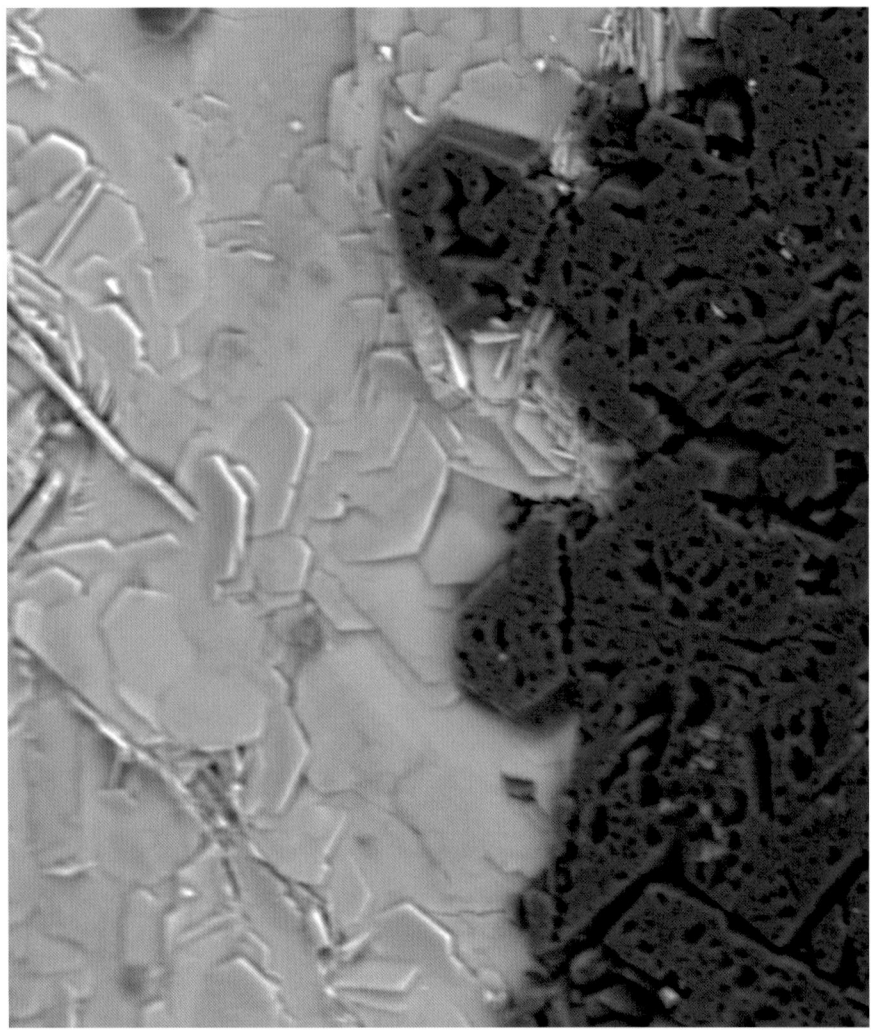

Rutile, TiO_2, occurs as microscopic inclusions in phlogopite.

Sphalerite, ZnS, rarely occurs as yellow, late-stage crystals to several mm.

Synchysite-(Ce), $Ca(Ce,La,Nd)(CO_3)_2F$, is found as late-stage, pseudohexagonal plates and elongated laths in fractures in tourmaline. The laths contain almost as much lanthanum as cerium.

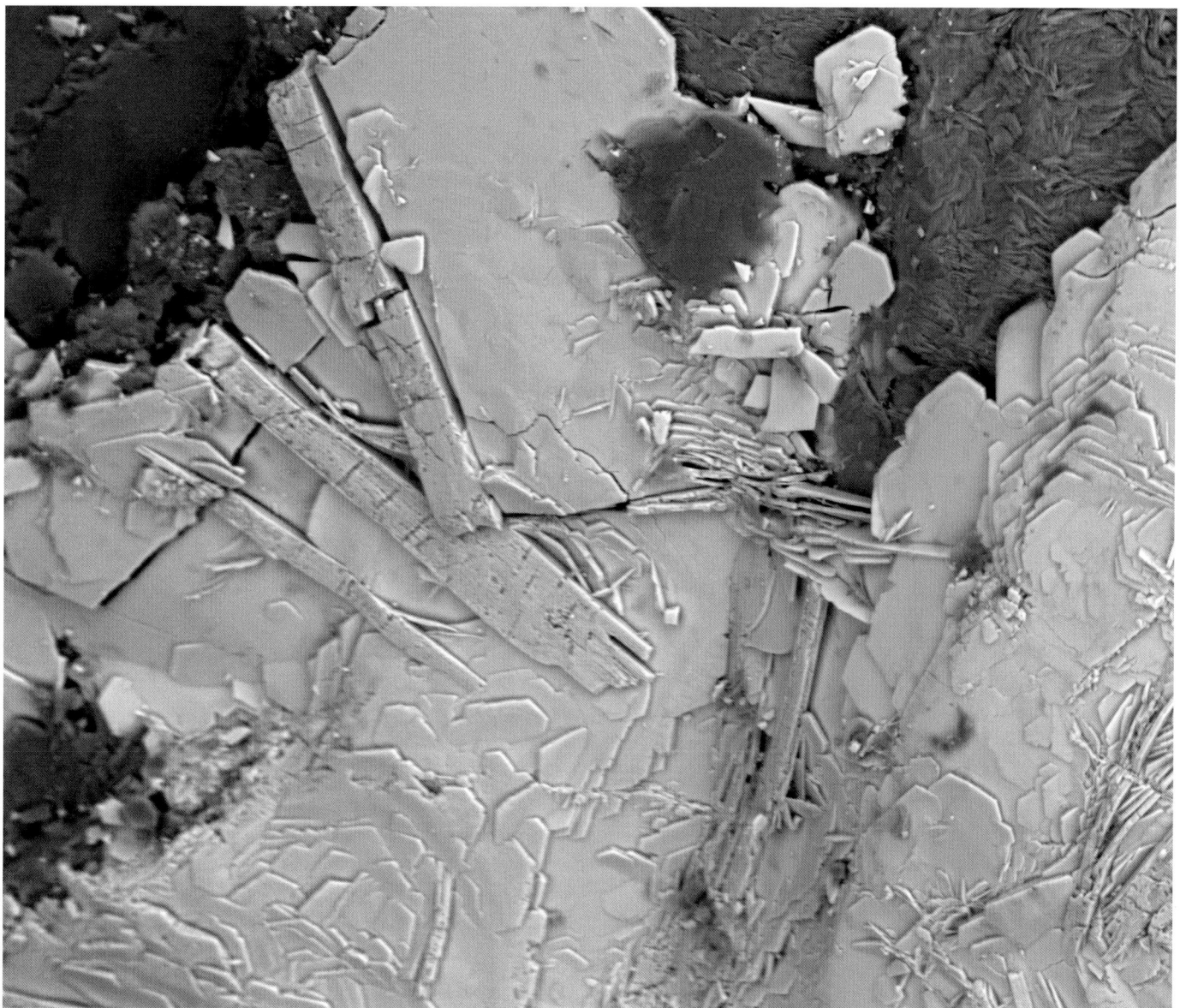

Figure 134. Synchysite-(Ce) and cordierite (?). 100 µm-fov. Chamberlain collection. *SCC & DGB*

Talc, $Mg_3(Si_4O_{10})(OH)_2$, occurs both as white to tan, massive coatings on quartz and other minerals, and as late-stage, transparent, micaceous crystals to several mm.

Figure 135. Talc. 13-mm fov. Chamberlain collection, #19222. *SCC*

Figure 136. Talc, chalcopyrite. 3.3-mm fov. Chamberlain collection, #19222. *SCC*

Figure 137. Talc, quartz, goethite. 14 cm. Collected by Mike Walter in 2008, now in the Chamberlain collection, #19118. *SCC*

Tourmaline is relatively abundant as black crystals in parts of the vein, but is often badly fractured and tends to break and crumble during cleaning. These fractures, however, contain late-stage crystals of synchysite-(Ce), quartz, and possible cordierite. Like the other tourmaline at the locality, the black tourmalines from this vein are dravite on the outside and fluor-uvite on the inside (See chapter 4).

Figure 138. Tourmaline. 5 cm. Collected by Mike Walter in 2008, now in the Chamberlain collection, #19199. SCC

Vermiculite, $Mg_{0.7}(Mg,Fe^{2+},Al)_6(Si,Al)_8O_{20}(OH)_4 \cdot 8H_2O$, is found as a brown, pearly alteration product of phlogopite.

Zircon, $ZrSiO_4$, occurs as crystal inclusions to a few microns in thin sections of black tourmaline.

Pseudomorphs

Chlorite ps. after phlogopite are typically only partially altered. Often on a cleavage surface the margins may be silvery or pearly green with the center still phlogopite. Sometimes the exposed end of a phlogopite crystal will be completely altered to chlorite. Often the chlorite species is clinochlore.

Clinochlore and muscovite ps. after marialite occur as dark, greenish-black prisms lining the vein wall. They are sharply formed and have a dull to somewhat shiny luster. Analysis with XRD indicates they are a mixture of clinochlore and muscovite. No unaltered marialite has been observed in the Phosphate Vein.

Figure 139. Clinochlore & muscovite after marialite, phlogopite, goethite. 5 cm. Collected by Steve Chamberlain in 2008 and in his collection, #18978. *SCC*

Goethite ps. after pyrite are relatively common as equant cubic crystals, sometimes flattened into square or rectangular wafers to several cm. Some of these have a small core of residual unaltered pyrite.

Figure 140. Goethite after pyrite (tabular). 4.4 cm. Collected by Steve Chamberlain in July, 2008, and in his collection, #19034. *SCC*

Quartz ps. after diopside look like prismatic diopside crystals to several cm that have been coated with transparent, granular sugar. Sometimes these pseudomorphs have only thick encrustations; others are completely replaced by fine-grained, white to transparent quartz.

Figure 141. Quartz after diopside. 3.3 cm. Collected by Mike Walter in 2007 and in his collection, #00104. *MW*

Quartz ps. after phlogopite occur as unusual, expanded phlogopite crystals with mm-thick plates separated by empty spaces and completely replaced by quartz. Late-stage quartz crystals are commonly found in the spaces between plates. The assemblage of expanded plates formed from one phlogopite crystal can be as long as 9 cm.

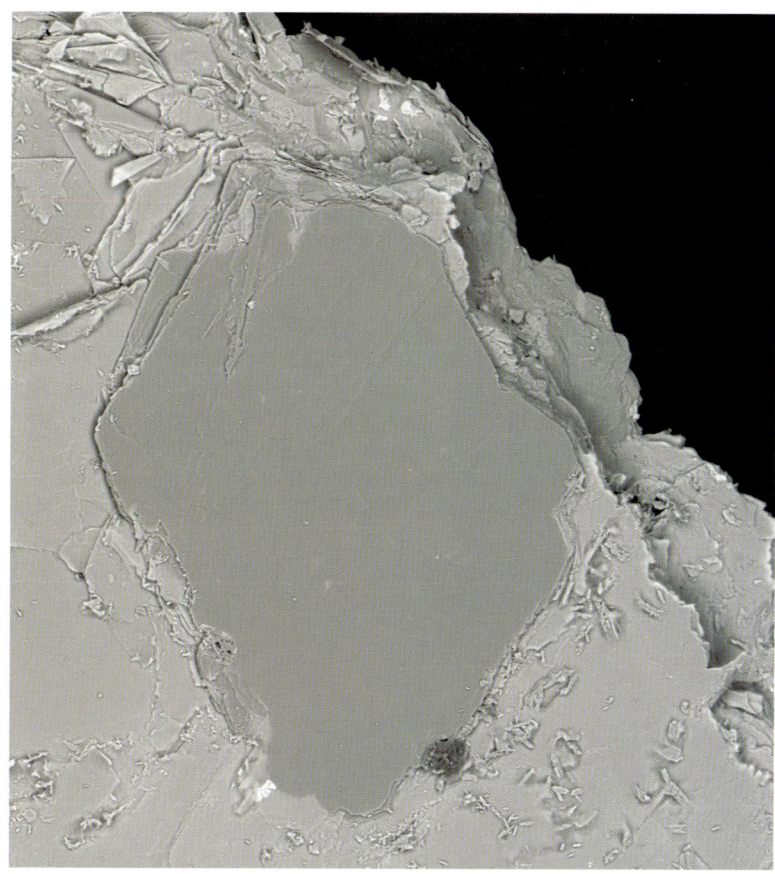

Figure 142. Quartz replacing phlogopite within cleavage sheet. 800-μm fov. Chamberlain collection. *SCC & DGB*

Figure 143. Quartz after phlogopite. 17-mm fov. Collected by Mike Walter in 2008, now in the Chamberlain collection, #19162. *SCC*

Figure 145. Quartz after phlogopite, Quartz. 6.4 cm. Collected by Mike Walter in 2008, now in the Chamberlain collection, #19161. SCC

Figure 144. Quartz after phlogopite. 4 cm. Collected by Mike Walter in 2008 and in his collection, #00059. MW

Figure 146. Quartz after phlogopite, quartz. 8.6 cm. Collected by Mike Walter in 2008, now in the Chamberlain collection, #19163. SCC

Figure 147. Quartz after phlogopite. 6.2 cm. Collected by Mike Walter in 2008 and in his collection, #00112. *MW*

Talc ps. after quartz occur as encrustation pseudomorphs after Tessin-habit quartz crystals that are thickly coated with white to tan massive talc. Most are only a few cm in length.

Vermiculite ps. after phlogopite are often only partially altered. The margins of a cleavage surface or the exposed end of a phlogopite crystal have a pearly brown appearance when altered to vermiculite.

Figure 148. Vermiculite after Phlogopite, 2 mm-fov. Chamberlain collection, #19564. *SCC*

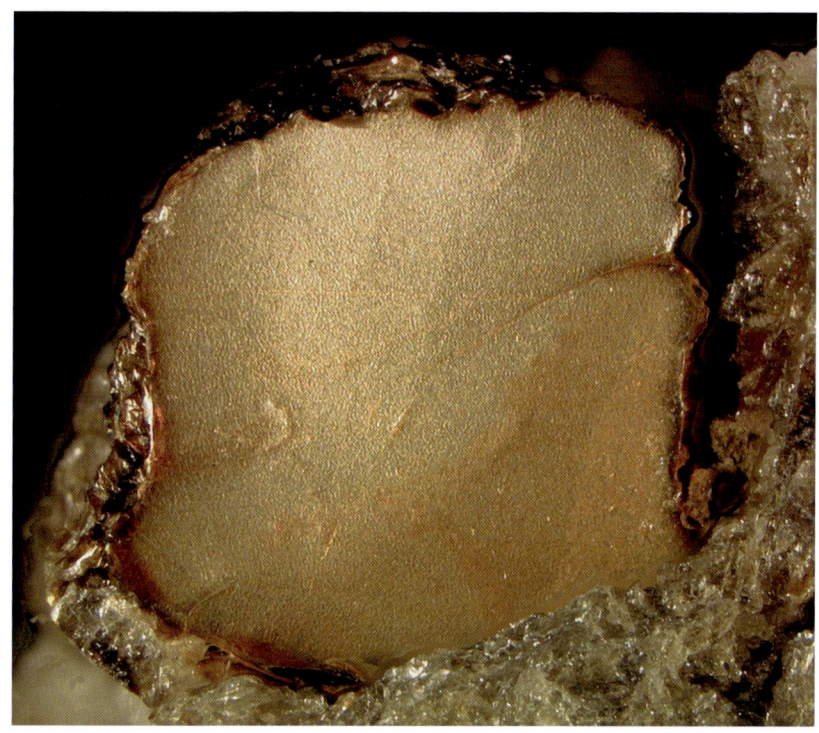

Middle Vein

The middle vein was fully visible as a narrow cleft in an outcropping of gneiss midway between the Waddell Vein and the Phosphate Vein. Only after extensive collecting at these two veins were excavations made on the Middle Vein late in 2008. The main minerals recovered were phlogopite, quartz, and tourmaline—all embedded in earthy goethite. Few specimens of interest to collectors were found.

Figure 149. Middle Trench as it appeared on October 23, 2008. *SCC*

Figure 150. Middle Trench as it appeared on June 12, 2014. *SCC*

Minerals

Chlorite, occurs as a silvery replacement of phlogopite.

Goethite, FeO(OH), is a common mineral as earthy masses and thick coatings on all the other minerals.

Fluorapatite, $Ca_5(PO_4)_3F$, occurs sparingly as tan, hexagonal prisms and basal pinacoids to 1 cm.

Phlogopite, $KMg_3AlSi_3O_{10}(OH)_2$, occurred as crystals several cm thick, but sometimes extremely elongated to 10 cm. Most of these are altered, at least in part, to vermiculite.

Pyrite, FeS_2, probably originally occurred in euhedral crystals, but extensive oxidative weathering has altered them to anhedral masses in goethite.

Quartz, SiO_2, occurs as Tessin-habit gray crystals to several cm.

Tourmaline is relatively uncommon, but a few clusters of cm-sized crystals to 14 cm were recovered. Like those found in the adjacent veins, these tourmalines have an outer rim of dravite and an inner core of fluor-uvite (see chapter 4).

Figure 151. Tourmaline. 9 cm. Collected by Mike Walter in the summer of 2009 and in his collection, #00079. *MW*

Vermiculite, $Mg_{0.7}(Mg,Fe^{2+},Al)_6(Si,Al)_8O_{20}(OH)_4\cdot 8H_2O$, is the principal alteration product of phlogopite.

Figure 152. Tourmaline. 14 cm. Collected by Mike Walter in August, 2009, now in the Chamberlain collection, #22969. SCC

Figure 153. Tourmaline, quartz. 4.2 cm. Collected by Mike Walter in summer 2009 and in his collection, #00087. MW

Pseudomorphs

Chlorite ps. after phlogopite tend to be silvery, rather than green, with a clear replacement texture. The alteration of phlogopite to chlorite appears to have caused the crystals to expand during replacement.

Figure 154. Chlorite after phlogopite. 6.5 cm. Collected by Donald Carlin Jr. in 2008, now in the Chamberlain collection, #19800. SCC

Vermiculite ps. after phlogopite preserve the original shape of the phlogopite crystals, but are satiny brown when cleaved. Many of the phlogopite crystals are completely altered to vermiculite.

Figure 155. Vermiculite after phlogopite. 9.8 cm. Collected by Mike Walter in the summer of 2008 and in his collection, #00047. MW

Waddell Vein

The Waddell Vein was discovered in the 1950s and intermittently worked by collectors until around 1968, after which it remained undisturbed until the spring of 2004. Serious collecting activities continued through 2008, by which time the vein had been excavated to the water level of the adjacent Leonard Brook and extended almost to the stream's edge. The mineralized portion of the Waddell Vein was significantly wider than the other three streamside veins and consisted of several parallel veins that variously split off from and rejoined the main vein.

Figure 156. Steve Chamberlain, Mike Hawkins, Brian Oveson, Scott Wallace, and Mike Walter at the Waddell Trench in 2005. *ML*

Figure 157. Waddell Trench as it appeared June 12, 2014. *SCC*

Minerals

Calcite, $CaCO_3$, originally formed the massive vein filling and was only encountered at the bottom of the excavation, the top portions having weathered away.

Chlorite, occurs as a green, pearly, nonelastic replacement of phlogopite. Clinochlore is one of the replacements of scapolite.

Goethite, $FeO(OH)$, occurs as a brown stain or coating on other minerals. Many of the pyrite crystals have a goethite coating of variable thickness.

Fluorapatite, $Ca_5(PO_4)_3F$, occurs sparsely as tan hexagonal prisms to several cm.

Mn-Fe oxides, a parallel vein immediately south of the main mineralized vein, contains large quantities of botryoidal dark brown to black aggregates of manganese-iron oxides. This material has not been characterized further.

Figure 158. Iron-manganese oxides. 16 cm. Collected by Scott Wallace in 2007, now in the Chamberlain collection, #23765. *SCC*

Phlogopite, $KMg_3AlSi_3O_{10}OH)_2$, occurs as dark brown to black pseudohexagonal prisms to 6 cm.

Figure 159. Phlogopite. 4.5 cm. Collected by Scott Wallace in 2008, now in the Walter collection, #00097. *MW*

Plagioclase, occurs in minute crystals coating quartz crystals. This feldspar was identified by XRD and is probably a late-stage mineral.

Figure 160. Plagioclase ps. quartz, Talc ps. Scapolite. 6 cm. Collected by Mike Walter in 2005 and in his collection, #00155. *MW*

Pyrite, FeS_2, occurs in cubic and octahedral crystals to 3.5 cm. While many are coated with goethite, some are not.

Figure 161. Pyrite. 3.2 cm crystal. Collected by Scott Wallace in August, 2004, now in the Chamberlain collection, #17992. *SCC*

Quartz, SiO_2, occurs as gray to white Tessin-habit crystals to several cm and also as transparent, colorless to white late-stage crystals to 3.5 cm.

Figure 162. Quartz. 2.5 cm. Collected by Scott Wallace in 2005, now in the Chamberlain collection, #18319. SCC

Talc, $Mg_3(Si_4O_{10})(OH)_2$, occurs as massive white to tan coatings, and as parallel arrays of transparent to translucent, white to gray, micaceous, late-stage crystals on quartz crystals.

Figure 163. Talc, Quartz. 13 mm. Chamberlain collection #19118. SCC

Tourmaline is abundant, often found in spectacular specimens. Equant, prismatic, and tabular crystals are common. Many of the prismatic crystals show striations on the prism faces, which is unusual for the locality. Clusters of tourmaline crystals to 15 cm have been recovered. Typically the crystals from this vein, and to some extent from the other three streamside veins, are more often tabular in habit than elsewhere at the locality. They are also more likely to have well developed prism striations, in sharp contrast to historical descriptions of these black tourmaline crystals as lacking striations. Crystals are dravite on the outside and fluor-uvite on the inside (see chapter 5).

Figure 164. Tourmaline. 5.8 cm. Collected by Mike Walter on July 27, 2004, now in the Chamberlain collection, #18845. *MW*

Figure 165. Tourmaline, 5.4 cm. Collected by Mike Walter and in his collection, #00071. *MW*

Figure 166. Tourmaline. 6.5 cm. Collected by Scott Wallace in 2004, now in the Chamberlain collection, #18297. *SCC*

Figure 167. Tourmaline. 6.7 cm. Collected by Mike Walter in 2005, now in the Chamberlain collection, #18855. SCC

Figure 168. Tourmaline, talc, quartz. 7 cm. Collected by Mike Walter, July 27, 2004. Chamberlain collection, #19118. MW

Figure 169. Tourmaline, talc. 6 cm. Collected by Mike Walter in summer 2005 and in his collection, #00073. MW

Vermiculite, $Mg_{0.7}(Mg,Fe^{2+},Al)_6(Si,Al)_8O_{20}(OH)_4 \cdot 8H_2O$, occurs as a brown, pearly, nonelastic replacement of phlogopite.

Pseudomorphs

Chlorite ps. after phlogopite are typically only partially altered. Often on a cleavage surface, the margins may be silvery or pearly green with a center of residual, unaltered phlogopite. Sometimes the exposed end of a phlogopite crystal will be completely altered to chlorite. Often the chlorite species is clinochlore.

Clinochlore and Muscovite ps. after marialite occur as shiny, opaque, dark green tetragonal prisms. Their mineral composition was determined by XRD.

Goethite ps. after pyrite are relatively common as both cubic and octahedral crystals. Many of these pseudomorphs are only altered on the surface.

Plagioclase ps. quartz forms dull, white coatings on Tessin-habit quartz crystals and resembles talc ps. after quartz.

Vermiculite ps. after phlogopite preserve the original shape of the phlogopite crystals, but are satiny brown when cleaved. Only rarely are the phlogopite crystals completely altered to vermiculite.

Wallace-Carlin Vein

The Wallace-Carlin Vein is the most northerly of the streamside veins, and was the one most recently discovered. In fall 2008, work on the Waddell Vein spilled over to a site several meters north where a collector had found some nice tourmaline in the soil. This led directly to the discovery of a fourth, parallel, vertical vein. Collecting continued through the fall of 2009. Compared to the other veins, the Wallace-Carlin Vein had a large amount of massive goethite from deep weathering of both pyrite and marcasite.

Figure 170. Wallace-Carlin Trench as it appeared on June 23, 2009. *SCC*

Figure 171. Tourmaline exposed on wall of Wallace-Carlin Trench, June 23, 2009. *SCC*

Figure 172. W-C Trench as it appeared on June 12, 2014. *SCC*

Minerals

Chlorite replaces many of the phlogopite crystals. Such replacements are pearly green and have inelastic cleavage plates.

Goethite, FeO(OH), is present as ocherous masses and coatings on the other minerals.

Fluorapatite, $Ca_5(PO_4)_3F$, is occasionally present as tan, hexagonal prisms to several cm.

Marcasite, FeS_2, although rare elsewhere across the locality, is common in this vein. Plates of cockscomb-twinned crystals to 2.5 cm, and hemispherical arrays to 10 cm, often coated with late-stage quartz, are relatively common. Much of the marcasite has weathered away to goethite, leaving crystal cavities.

Figure 173. Marcasite. 2.5 cm. Collected by Scott Wallace in summer, 2009, now in the Walter collection, #00148. *MW*

Figure 174. Marcasite. 15 mm fov. Chamberlain collection, #20020. *SCC*

Phlogopite, $KMg_3AlSi_3O_{10}(OH)_2$, is found as pseudohexagonal crystals, frequently wholly or partially replaced by either chlorite or vermiculite.

Pyrite, FeS_2, occurs as shiny, gold-colored crystals of relatively complex habit to 3 cm or more. Larger anhedral blobs of pyrite thickly covered with goethite are also common.

Figure 175. Pyrite. 1.3 cm. Collected by Scott Wallace in 2008, now in the Walter collection, #00149. MW

Figure 176. Pyrite. 3.1 cm. Collected by Scott Wallace in September, 2008, now in the Chamberlain collection, #29451. SCC

Quartz, SiO_2, occurs sparingly as Tessin-habit crystals and more commonly as late-stage, transparent crystals, usually smaller than 1 cm.

Figure 177. Quartz. 3.9 cm. Collected by Donald Carlin Jr. in 2008, now in the Chamberlain collection, #19731. SCC

Tourmaline occurs as crystals to 8 cm and groups to 15 cm or larger. Tabular crystals and those with odd shapes due to differential growth seem to be more common than elsewhere on the locality. The tourmaline is dravite on the surface and fluor-uvite on the inside (see chapter 5).

Figure 178. Tourmaline. 8.5 cm. Collected by Scott Wallace in Spring 2009, now in the Walter collection, #00137. *MW*

Figure 179. Tourmaline. 6.3 cm. Collected by Scott Wallace in Fall 2008, now in the Walter collection, #00038. *MW*

Figure 180. Tourmaline, quartz. 6.5 cm. Collected by Mark Norman, Mike Bicknell, Cody Meashaw, Brett Miller, and Ryan Gallagher in 2008, now in the Chamberlain collection, #19714. *SCC*

Figure 181. Tourmaline, goethite. 8.5-cm crystal. Collected by Donald Carlin Jr. in 2008, now in the Chamberlain collection, #18732. *SCC*

Vermiculite replaces many of the phlogopite crystals. Such replacements are pearly brown and have inelastic cleavage plates.

Pseudomorphs

Chlorite ps. after phlogopite preserve the original shape of the phlogopite crystals, but are satiny green when cleaved. Many of the original phlogopite crystals are largely altered to chlorite.

Vermiculite ps. after phlogopite preserve the original shape of the phlogopite crystals, but are satiny brown when cleaved. Many of the original phlogopite crystals are largely altered to vermiculite.

Hillside

The hillside site is southeast of the classic site on top of the hill. Exploration by local collectors revealed this mineralized area spread over a grass-covered hill in the 1950s and early 1960s. Both outcropping mineralization and mineralized boulders are scattered across the area.

The hillside is usually reached from the entry pasture by bearing to the right (straight) along an old farm lane. After a distance of several hundred meters, the land begins to rise on the right, and turning right and walking uphill leads to the hillside site.

Figure 182. Hillside site as it appeared in Summer 1983. *Fred Ruhe photo*

Minerals

Diopside, $CaMgSi_2O_6$, formed as blocky crystals to 5 cm; however, most of these have been replaced by quartz or tremolite and are now pseudomorphs.

Phlogopite, $KMg_3(AlSi_3O_{10})(OH,F)_2$, occurs sparingly as pseudohexagonal black prisms, often partially altered to chlorite

Quartz, SiO_2, is rare as crystals, but common as a granular, white replacement of diopside crystals.

Tourmaline occurs as crystals to 8 cm and groups to 30 cm or larger. Tabular crystals and those with odd shapes due to differential growth seem to be more common than elsewhere on the locality. The tourmaline is dravite on the surface and fluor-uvite on the inside (see chapter 5).

Figure 183. Tourmaline. 5 cm. Collected by Mike Walter in Summer 1993, now in Jay Walter collection. *MW*

Figure 184. Tourmaline. 7 cm. Collected by Elmer Rowley in 1965, now in the New York State Museum collection, #10366. *Stephen Nightingale photo*

Chapter 3: The Sites and Their Minerals

Figure 185. Tourmaline. 3.6 cm. Collected by Schuyler Alverson in 1950, now in the Chamberlain collection, #4257. *SCC*

Figure 186. Tourmaline, quartz, Chlorite after Phlogophite. 14.3 cm. Collected by Ed Nesbitt in 1964, now in the Chamberlain collection, #17867. *SCC*

Figure 187. Tourmaline. 9.8 cm. Collected by Schuyler Alverson in 1979, now in the Chamberlain collection, #5176. *SCC*

Tremolite, $Ca_2Mg_5Si_8O_{22}(OH)_2$ to $Ca_2Mg_{4.5}Fe^{2+}_{0.5}Si_8O_{22}(OH)_2$, frequently forms epitactic overgrowths of diopside. Most often the underlying diopside has been replaced by granular, white quartz.

Pseudomorphs

Chlorite ps. after phlogopite constitute most of what first appears to be phlogopite crystals at this site. On larger crystals, the alteration is only a zone on the outside.

Quartz ps. after diopside are common here, as quartz replaces almost all the diopside. The quartz forms a white, granular replacement structure that preserves the blocky shape of the original diopside. Tremolite epitactic on the original diopside remains as unaltered, dark green parallel growths of crystals.

Figure 188. Quartz & tremolite after diopside. 2.8-cm crystal. Collected by Steve Chamberlain and in his collection, #7891. *SCC*

Tremolite ps. after diopside are dark green, blocky crystals where epitactic tremolite covered the surfaces of diopside crystals. Often the internal diopside has been completely replaced by granular, white quartz.

Marsh

The marsh, also referred to by collectors as the "swamp," is a boulder field in a marshy area along the southeastern edge of the hillside site. It is usually reached by walking across the hillside site from the farm lane and then climbing down the steep bank. Mostly very large specimens have been recovered from this site where the calcite is largely weathered away, exposing the crystallized minerals.

Figure 189. Marsh (swamp) site as it appeared in June, 2014. *JC*

Minerals

Phlogopite, $KMg_3AlSi_3O_{10}(OH)_2$, occurs as pseudohexagonal, black prisms to several cm with quartz and dravite.

Quartz, SiO_2, is found as gray, Tessin-habit crystals to several cm and as anhedral, gray masses.

Tourmaline occurs as arrays of equant, black crystals on matrix and as crystals with quartz and phlogopite. The crystals are dravite on the outside and fluor-uvite inside (see chapter 5).

Figure 190. Tourmaline. 18 cm. Collected by Ron Waddell in 1961, now in the Chamberlain collection, #15891. *SCC*

Figure 191. Tourmaline, quartz. 45 cm. Collected by Dick Stimer in1960, now in the New York State Museum collection, #21153, *Stephen Nightingale photo*

CHAPTER 4

GALLERY

Having described and illustrated the minerals that occur at each of the sites across the Pierrepont black tourmaline occurrence, we now present a gallery of some particularly excellent specimens of black tourmaline from museum and private collections. These specimens indicate the importance of this classic locality.

Figure 192. Tourmaline. 12.5 cm. Top of Hill site. In the W. E. Hidden collection at the Natural History Museum, Vienna, Austria, #E5232. *Alice Schumacher photo*

Figure 193. Tourmaline. 10 cm. Top of Hill site. Collected by Mike Walter in 1994, now in the Hollmann collection at the New York State Museum, #24307. *Ken Hollmann photo*

Figure 194. Tourmaline, quartz. 7 cm. Top of Hill. Collected by George Robinson in 1968, now in the Canadian Museum of Nature collection, #51173. *GWR*

Figure 195. Tourmaline, diopside. 5.2-cm fov. Top of Hill. In the George Brush collection at Yale University, #025139. *Fred E. Davis photo*

Figure 196. Tourmaline. 8.4 cm. Top of Hill. Collected by C. D. Nims, 1890, now in the Chamberlain collection, #18851. *SCC*

Figure 197. Tourmaline. 15 cm. Wallace-Carlin Vein. Collected by Scott Wallace and Donald Carlin Jr. in Spring 2009, now in the Mike Walter collection, #0009. MW

Figure 198. Tourmaline. 11.5 cm. Top of Hill. From the Harvard University collection, now in the Chamberlain collection, #14161. SCC

Figure 199. Tourmaline. 6.8 cm. Waddell Vein. Collected by Mike Walter on July 27, 2004, now in the Chamberlain collection, #15462. MW

Chapter 4: Gallery

Figure 200. Tourmaline, quartz. 19 cm. Top of Hill. Collected by Vern Phillips in 1985, now in the Chamberlain collection, #9125. *SCC*

Figure 201. Tourmaline, calcite. 15 cm xl. Top of Hill. Collected by Dale A. Russell in 2011 and in his collection. *Catherine Kozaczka photo*

113

CHAPTER 5

SPECIAL FOCUS TOPICS

Tourmaline—Composition and Nomenclature
At the time of the locality's discovery in 1859 by teenager R. T. Cross and the subsequent worldwide distribution of its black tourmaline specimens by C. D. Nims over the next several decades, tourmaline was considered a species and the varieties were usually classified by color. By the nomenclature of the sixth edition of *Dana's System of Mineralogy* (1892), the black tourmaline from Pierrepont would have been categorized as "normal." At the time, schorl was considered a European, chiefly German, term for opaque black "normal" tourmaline, although it had been in use since 1564 (Dietrich, 1985).

Once composition became an important classification tool and tourmaline became a group name, most opaque black tourmaline was assumed to be schorl. Specimens of black tourmaline from Pierrepont are still sometimes labeled "schorl," even though this is not even remotely correct.

In 1977, Dunn and his colleagues published a paper establishing uvite as a member of the tourmaline group (Dunn et al., 1977). They suggested that the black tourmaline from Pierrepont be labeled ferroan (or iron-rich) uvite. Since then, specimen mineralogists have tended to label Pierrepont tourmaline as uvite. Researchers, by contrast, often called it dravite (e.g. Grice and Ercit, 1993; Dyar et al., 1998). With the description of the hydroxyl end member as uvite (Clark et al, 2010), uvite containing dominant fluorine, such as the black tourmaline from Pierrepont, became fluor-uvite in the parlance of specimen mineralogists.

With the publication of the new nomenclature for the tourmaline supergroup (Henry et al., 2011) along with rules for converting analyses into formulae, the correct nomenclature for the black tourmaline from Pierrepont became dravite (Lupulescu and Hawkins, 2013; Chamberlain and Robinson, 2013). Under the new nomenclature rules, the composition of the anions at the end of the formula, usually either OH or F, must also be considered to determine the species. Thus, uvite became two species: uvite (with dominant OH) and fluor-uvite (with dominant F). Likewise, dravite became dravite (with dominant OH) and fluor-dravite (with dominant F). Although some of the historical confusion about the species identity of the Pierrepont black tourmaline—schorl, uvite, fluor-uvite, dravite—resulted from the evolving nomenclature and technology for quantifying fluorine in silicates, one principal question remained. Some modern analyses did not indicate enough fluorine to make the Pierrepont tourmaline either fluor-dravite or fluor-uvite. These analyses gave a composition of dravite. Other modern analyses showed sufficient fluorine to make it either fluor-dravite or fluor-uvite, but with Ca>Na, yield compositions within the field of fluor-uvite.

Our examination of the composition of the black tourmaline from Pierrepont, detailed below, suggests a possible explanation for this problem. During the main-stage emplacement of the pegmatite across the locality, the crystallization of tourmaline and other minerals (principally quartz, diopside, and phlogopite) may have caused the relative depletion of some components in the liquid phase (such as calcium and fluorine) and a relative increase in others (such as sodium and hydroxyl ion). Until a thorough survey is done of the full composition of tourmaline from across the occurrence, the working hypothesis is that the pattern of zoned composition may well apply to all the black tourmaline from the locality, i.e. the interior is fluor-uvite; the outer rind is dravite.

Optical Microscopy. A four-centimeter tourmaline crystal was sectioned parallel to the *c*-axis and two polished thin sections were made and examined with transmitted polarized light using a petrographic microscope. Under this illumination, Pierrepont tourmaline displays light zoning, fractures, and inclusions of calcite (dominant), phlogopite, rounded grains of tourmaline from another generation, pyrite partially replaced by goethite, and quartz. The calcite seems to follow a "paleo-surface" of the crystal in one of the thin sections. This could be interpreted as a hiatus in the growth of the crystal, with the incorporation of minor calcite from the host vein, followed by continued growth of the tourmaline. The crystal also has an intermittent narrow rim displaying more intense pleochroism.

Figure 202. Thin section cut along the *c*-axis of a tourmaline crystal from Pierrepont. Note darker rim. *ML*

Figure 203. Higher magnification of darker external rim in thin section. *ML*

Figure 204. Oriented zoning in the core of a tourmaline crystal with quartz inclusions. *ML*

Electron Microprobe Analysis. One of the thin sections was carbon coated under vacuum for analysis by electron microprobe with a Cameca SX-100 in the Microanalytical Laboratory, Department of Geology, Rensselaer Polytechnic Institute, Troy, New York. The operating conditions were: accelerating voltage 15 kV, beam current 20 nA and beam diameter 5 μm. Standards used were jadeite (Na), orthoclase (K), diopside (Ca), synthetic fayalite (Fe), synthetic forsterite (Mg), tephroite (Mn), kyanite (Al, Si), rutile (Ti), topaz (F), and sodalite (Cl). The data were reduced using a standard ZAF correction routine.

Major elements were determined for twenty-one points along the "c" axis of the crystal and one point on its rim (Table V1). H_2O and B_2O_3 were calculated by stoichiometry, assuming (OH + F) = 4 apfu and B = 3 apfu.

The resulting empirical formula for the average of the twenty-one analyses along the "c" axis of the crystal is: $(Ca_{0.62}Na_{0.37})_{S1.00}(Mg_{2.01}Fe_{1.00}Ti_{0.07})_{S3.08}(Al_{5.11}Mg_{0.89})_{S6.00}(Si_{5.97}Al_{0.03}O_{18})(BO_3)_3(OH)_{3.00}(F_{0.80}OH_{0.19})$, which is the species fluor-uvite.

The empirical formula for the analysis of the rim of the crystal is: $(Ca_{0.49}Na_{0.54})_{S0.99}(Mg_{1.76}Fe_{1.23}Ti_{0.12})_{S3.11}(Al_{5.16}Mg_{0.94})_{S6.00}(Si_{5.92}Al_{0.08}O_{18})(BO_3)_3(OH)_{3.00}(OH_{0.64}F_{0.36})$, which is the species dravite.

The concentrations of major elements determined by electron microprobe were very consistent along the "c" axis with very small variations in Fe (range 6.80 to 7.40 wt %), which might account for the light zoning seen in thin section, while the darker appearance and more intense pleochroism observed in the rim might be due to a combination of slightly greater concentrations of Fe and Ti.

Laser ablation-inductively coupled plasma-mass spectrometry. The trace elements were determined by LA-ICP-MS in a polished thin section using a detachable Photon Machines Analyte, a 193 ultra-short pulse excimer laser ablation (LA) workstation, and a Varian 820-MS inductively coupled plasma mass spectrometer (ICP-MS). These measurements were made in the same laboratory as the electron microprobe analyses

Trace elements were analyzed in two subgroups according to their atomic numbers and then combined as shown in Table V2. Vanadium, mercury, and strontium had the highest values, followed by zinc, chromium, and gallium. Among the rare-earth elements, cerium had the highest value followed by lanthanum. All other analyzed trace elements were at very low concentrations.

OXIDE	BP	Rim
SiO_2	36.04	35.42
Al_2O_3	26.36	26.55
TiO_2	0.55	0.97
FeO	7.22	8.76
MgO	11.78	10.45
CaO	3.48	2.50
MnO	0.02	0.01
Na_2O	1.17	1.67
K_2O	0.05	0.07
F	1.54	0.69
Cl	0.01	0.00
H_2O*	2.89	3.26
B_2O_3*	10.50	10.39
Total	101.60	100.73
O=F	0.65	0.29
Total	100.96	100.44
apfu		
Si^{4+}	5.97	5.92
Al^{3+}	0.03	0.08
ΣT	6.00	6.00
B	3.00	3.00
Al^{3+}	5.11	5.16
Mg^{2+}	0.89	0.84
ΣZ	6.00	6.00
Ti^{4+}	0.07	0.12
Mg^{2+}	2.01	1.76
Fe^{2+}	1.00	1.23
ΣY	3.08	3.11
Ca^{2+}	0.62	0.45
Na^+	0.37	0.54
ΣX	0.99	0.99
OH	3.19	3.64
F	0.80	0.36

Figure 205. Table of major elements from electron microprobe analysis for the center (BP) and the rim (Rim) of black tourmaline from Pierrepont.

Zn	64.33	Dy	0.01	V	316.8	Nb	0.13
Ga	32.47	Er	0.01	Cr	45.78	Cd	0.08
La	9.67	Yb	0.02	Co	4.45	In	0.04
Ce	11.90	Lu	0.01	Ni	4.52	Sn	2.03
Pr	0.77	Hf	0.03	Cu	0.28	Sb	2.02
Nd	1.70	Ta	0.03	Ge	0.56	Ba	2.29
Sm	0.05	Hg	114	Sr	631.2		
Eu	0.22	Pb	0.76	Y	0.04		
Gd	0.03	Sc	8.89	Zr	0.36		

Figure 206. Table of trace elements in parts per million (ppm) from laser ablation-inductively coupled plasma-mass spectrometry for the center of a black tourmaline from Pierrepont.

Proper Nomenclature for Pierrepont Black Tourmaline. There are many thousands of specimens of the lustrous black tourmaline crystals from Pierrepont in the mineral collections of the world. How should they be labeled?

Three issues cloud this determination. 1) Until now, no one knew that at least some of the crystals were concentrically zoned with dravite on the outside and fluor-uvite in the center. Therefore, samples from different parts of a crystal will give different compositions. 2) Being an artificial construct, the nomenclature happens to put the boundary between dravite and uvite right in the middle of the cloud of data points arising from multiple analyses of unselected samples. Thus, small differences can change the determined species. 3) Tourmaline crystals have not been exhaustively examined along the length of any of the veins nor across the sites at the locality. While it is likely that the pegmatite that gave rise to the tourmaline might not have varied much over these distances, this possibility has not been evaluated.

Specimen mineralogists tend to use one good analysis as the identification for all samples of the same sort from a locality. Using that traditional approach, complete crystals of black tourmaline from Pierrepont are dravite on the surface. Broken fragments are fluor-uvite. However, the compositional differences between the interior core and exterior rim of these crystals are so small that making these distinctions is hardly useful. Specimens already labeled "dravite" can remain thus. Specimens already labeled "uvite" should be updated to "fluor-uvite" now that uvite has been divided into uvite, with dominant OH, and fluor-uvite, with dominant F. *We therefore recommend that unanalyzed specimens of black tourmaline from Pierrepont simply be labeled "tourmaline," as we have done in this volume.*

Tourmaline—Crystal Forms and Habits
Crystal Forms

Tourmaline crystallizes in the ditrigonal pyramidal (3m) class of the hexagonal system. In this class, all the crystal forms are open so that a crystal must have at least two, and usually three different forms to enclose space. Because there is no center of symmetry, the crystal structure of tourmaline crystals is hemimorphic. This may be reflected in the crystal morphology so that the two terminations have different appearance (see Rakovan, 2007, for a thorough discussion). The forms in the prism zone parallel to the c-axis include trigonal prisms, the hexagonal prism, and ditrigonal prisms. The forms making up the termination include the pedion, trigonal pyramids, the hexagonal pyramid, and ditrigonal pyramids.

Figure 208. Tourmaline with hemimorphic development in calcite. The steep faces on the top are the trigonal pyramid {021}; those on the bottom, the flatter trigonal pyramid {01-1}. 2.4-cm crystal. Top of Hill. Collected by Mike Walter in 2014, now in the Chamberlain collection, #31536. *SCC*

Figure 207. Broken tourmaline group showing the outer zone of dravite (red) and the inner core of fluor-uvite (blue). Our measurements were not comprehensive across the locality, but since the tourmaline crystals formed from a pegmatite it is likely all are zoned in this manner. *Painting by Susan Robinson*

The black tourmaline crystals at Pierrepont usually show a subset of the following dominant forms: the trigonal prism m{100}, the ditrigonal prism a{110}, the steeper trigonal pyramid o{021}, and the flatter trigonal pyramids r{101}, z{011}, and e{012}. More rarely the pedion c{001}, the steeper trigonal pyramid y{401}, or the steeper ditrigonal pyramids t{211} and u{321} also may be present.

Because tourmaline has no center of symmetry, the identity of a form on the top (antilogous pole) and the same form on the bottom (analogous pole) of a given crystal will have a positive c-axis value or a negative c-axis value respectively. So a pedion on top will be {001}, but {00-1} on the bottom.

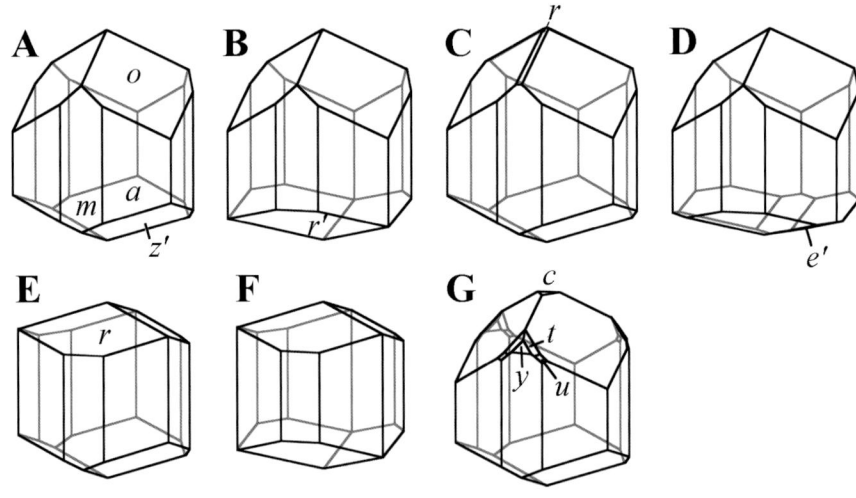

Figure 209. Crystal drawings showing the crystal forms of Pierrepont tourmaline. A shows hemimorphic development with the steep trigonal pyramid, o{021}, above and the shallow trigonal pyramid, z'{01-1}, below. The middle consists of the hexagonal prism, m{100}, and a ditrigonal prism, a{110}. B shows the same forms as A except that a different shallow trigonal prism, r'{10-1}, appears below. C is the same as A with the additional modifying shallow trigonal prism, r{101} above. D is the same as B with the additional modifying shallow trigonal prism, e'{01-2} below. E shows the shallow trigonal pyramid, r{101}, above and the shallow trigonal pyramid, z'{01-1}, below. F is the same as E, but with a different shallow trigonal pyramid, r'{10-1} below. G shows the drawing in A with minor modifying faces: pedion, c{001} and steep trigonal pyramids, t{211}, u{321}, and y{401}.

The crystal drawings illustrate some of the most commonly found combinations of forms. A strongly hemimorphic crystal will have the steep trigonal pyramid o{021} on the top and either the flat trigonal pyramid z'{01-1} (fig. 209, A) or the flat trigonal pyramid r'{10-1} (fig. 209, B) on the bottom. Sometimes the top trigonal pyramid will be modified by small faces of the trigonal pyramid r{101} (fig. 209, C). Similarly, the bottom trigonal pyramid r'{10-1} will be modified by small faces of the trigonal pyramid e'{01-2} (fig. 209, D).

If both the top and bottom have flat trigonal pyramids, the crystal does not look hemimorphic and is often much harder to orient. Two such situations are illustrated. If the top trigonal pyramid is r{101} and the bottom trigonal pyramid is z'{01-1} (fig. 209, E), the two pyramids appear rotated about the c-axis. If the top trigonal pyramid is r{101} and the bottom trigonal pyramid is r'10-1} (fig. 209, F), the two pyramids appear to be in alignment.

Faces from a steeper trigonal pyramid and two steeper ditrigonal pyramids are usually minor (fig. 209, G). These are occasionally present, although usually not together on one crystal.

For tabular crystals discussed below, the arrangement of forms is the same, but the prism zone is shortened along the c-axis into a more-or-less narrow girdle.

Crystal Habits. Habit refers to the overall appearance of a crystal and is a qualitative characteristic because it typically depends on observation and not measurement. There are four principal habits of black tourmaline crystals from Pierrepont: equant, tabular, prismatic, and distorted.

Most of the black tourmaline crystals at Pierrepont are *equant*—those whose c-axis length is the same as the diameter of the crystal in the plane of the a-axes. These crystals fit nicely into a sphere. For equant black tourmaline crystals from Pierrepont, the faces of the dominant forms are often about the same size. Moreover, the various prism faces almost always lack the striations generally found on tourmaline crystals from many localities. Together these features make determining the crystallographic orientation of individual crystals in aggregates somewhat difficult. Fortunately, complete tourmaline crystals that formed isolated in calcite ("floaters") are relatively common, so that crystal habit can be studied from these complete, isolated crystals.

Tabular crystals are the next most common habit. They have the same combinations of crystal forms as outlined above, but the trigonal and ditrigonal prism faces are shortened along the c-axis so that the crystals are disc-shaped with a narrow prism zone.

Prismatic crystals are the least common habit. In contrast to tabular crystals, they are elongated along the c-axis, with a length:diameter ratio from 1.5 to 1.75.

Distorted crystals are all those that don't fit in the above three habits. Some show selective growth of some faces of one form over other faces of that form, resulting, for example, in crystals flattened along one a-axis, but not the other two. Such preferred growth also causes selectively fast-growing faces of some forms to disappear so that crystals have acutely pointed corners. Some of these growth-distorted tourmaline crystals require serious crystallographic measurements to interpret. Skeletal crystals, formed when edges grow faster than the centers of faces, are occasionally encountered, but are relatively rare. Likewise, examples of extensive parallel growth are occasionally found.

To estimate the approximate distribution of these four categories, we examined 571 single-crystal thumbnail specimens of black tourmaline and grouped them into the categories of equant, tabular, prismatic, and distorted.

Figure 210. Skeletal tourmaline. 2.8 cm. Top of Hill. Collected by Donald Briggs in 1985, now in the Chamberlain collection, #22054. SCC

Figure 211. Tourmaline, quartz. 9-cm fov. Top of Hill. Parallel growth. Collected by Steve Chamberlain in July, 1985 and in his collection, #6560. SCC

Figure 212. Tourmaline, calcite. 22 cm. Top of Hill. Parallel growth. Collected by James Dee in 1988, now in the Chamberlain collection, #11647. SCC

Categorization into one of the four groups was done by visual examination. The results of this analysis are: equant - 75 percent; tabular - 18 percent; prismatic - 2 percent; distorted - 5 percent.

Habits of Quartz

Quartz at Pierrepont exhibits three primary crystal habits: tapered crystals with more or less prominent striations on the prism faces perpendicular to the c-axis—the so-called Tessin habit; elongated crystals with planar, vitreous, prism faces and terminal faces made up of positive and negative rhombohedra; and equant crystals with minor vitreous prism faces and prominent terminal faces made up of rhombohedra. Tessin quartz forms at high temperature. The other two habits are typical of low-temperature hydrothermal mineralization.

Figure 213. Principal habits of quartz at Pierrepont. The Tessin habit on the left is characteristic of the primary pegmatite mineralization. The normal habit on the right is characteristic of later, low-temperature hydrothermal mineralization. The equant crystal in the middle is an uncommon habit in the low-temperature hydrothermal mineralization. *Sketches by Susan Robinson*

It is thought that Tessin habit quartz typically forms at temperatures between 500 degrees and 600 degrees C, often from fluids that are charged with CO_2 (Mullis et al. 1994; Stalder and Weisbrod, 1974). At Pierrepont, Tessin habit crystals crystallized contemporaneously with tourmaline and phlogopite during the main-stage emplacement of the pegmatite, forming crystals that are generally strongly tapered, sometimes smoky, often only weakly striated, and characterized by rounded edges between adjacent crystal faces.

Quartz of "normal habit," with sharp, parallel, prism faces, few striations, and well-defined rhombohedral terminations probably formed from lower temperature hydrothermal solutions. Fluid-inclusion studies of such veins in the region yield formation temperatures between 150 degrees and 175 degrees C, although we have not examined fluid inclusions from the quartz of this habit from Pierrepont. The normal-habit quartz at Pierrepont is a late-stage mineral. It varies from colorless to white, and from pale to dark smoky color. Striations are infrequent, although large domains of slightly different growth features are common on prism faces. The terminations vary from symmetrically equivalent development of the positive and negative rhombohedra (resulting in a pseudohexagonal dipyramidal habit), to three larger rhombohedral faces alternating with three smaller faces (trigonal habit), to one large rhombohedral face and five much smaller such faces (Dauphiné habit).

Equant quartz crystals, sometimes called Cumberland-habit quartz, are relatively rare at Pierrepont and occur with those of normal habit in late-stage, presumably low-temperature hydrothermal veins. These have prominent positive and negative rhombohedral faces and almost no visible prism faces. Both pseudohexagonal and trigonal habits of equant quartz crystals occur. They are usually colorless to white to mottled dark brown.

Selected Pseudomorphs

Numerous pseudomorphs found at the various sites comprising this locality are listed and described in chapter 3. These pseudomorphs represent two of the major types: pseudomorphs by encrustation, and pseudomorphs by alteration. In some cases these two types overlap. For example, a quartz crystal can become encrusted with white talc and also be partially altered to talc within the original quartz crystal. Most of the pseudomorphs occurring at this locality are relatively common at localities across the Grenville Geological Province in New York, Ontario, and Quebec. Many of them formed during a period of retrograde metamorphism after the high temperatures and pressures of regional metamorphism had peaked. Because they were no longer stable at the lower temperatures and pressures, some of the previously formed, high-temperature minerals were changed into other minerals that were more stable at the lower temperature/pressure conditions. Others represent alteration by chemical weathering, such as goethite ps. after pyrite. Some of these pseudomorphs may have formed during the period of late-stage hydrothermal mineralization that deposited quartz, talc, and other silicates along with rare-earth minerals in open spaces.

Figure 214. Goethite after pyrite. 3.9 cm. Top of Hill. Collected by Stephen Miller in June 2008, and in his collection. *SCC*

Four categories of pseudomorphs seem to merit additional discussion: the alteration of phlogopite, the replacement of pyrrhotite by pyrite, the alteration of diopside, and the alteration of scapolite. Some of these pseudomorphs are unusual, and, like all pseudomorphs, provide direct evidence of changing geological conditions. Through their study, these pseudomorphs may provide important clues that help unravel some of the complex geological history that is responsible for the origin of this most interesting locality.

Vermiculite ps. after Phlogopite and Chlorite ps. after Phlogopite. Iron-rich phlogopite is relatively abundant as one of the minerals that crystallized from the original Precambrian pegmatite. Many specimens of fresh, unaltered phlogopite collected at the various sites across the locality have shiny, transparent, dark brown cleavage surfaces. However, some specimens have pearly, translucent, light brown areas on their cleavage planes. Sometimes the edges of a cleavage plate are light brown and the center is still transparent brown phlogopite. Other specimens show a pearly, translucent green appearance on cleavage planes, again sometimes with transparent, brown phlogopite centers. These pearly brown and green areas represent the alteration of phlogopite to two other similar sheet silicates. We have not exhaustively analyzed altered phlogopite specimens from all the sites across the locality, but the pattern is clear. The light brown is typically vermiculite; the green is often clinochlore.

These three minerals are monoclinic sheet silicates with perfect basal cleavage and similar compositions: phlogopite - $K(Mg,Fe)_3(AlSi_3O_{10})(OH,F)_2$; clinochlore - $(Mg,Fe)_5Al(AlSi_3O_{10})(OH)_8$; vermiculite - $(Mg,Fe,Al)_3((Al,Si)_4O_{10})(OH)_2 \cdot 4H_2O$. As a true mica, unaltered, thin cleavage sheets of phlogopite are elastic, snapping back into position when deformed. Clinochlore (a chlorite) and vermiculite (a saponite group mineral) have cleavage sheets that bend, but then remain bent (i.e, flexible, but inelastic).

The alteration of iron-rich, brown phlogopite to pearly light brown vermiculite is fairly well understood (e.g. Toksoy-Koksal et al., 2001). Alteration proceeds along the cleavage planes and is coherent, i.e. does not cause a significant change in volume. As a consequence, no sharp alteration boundaries are observed between a region of vermiculite and unaltered phlogopite in a cleavage sheet. Vermiculite ps. after phlogopite from Pierrepont preserve the external appearance of the original phlogopite crystal and usually do not show any expansive separation of the sheets along cleavage planes.

The alteration of iron-rich brown phlogopite to chlorite is more complex (Yau et al., 1984). One mechanism is coherent and does not cause a significant change in volume. Like the alteration to vermiculite, no sharp alteration boundary is visible between the chlorite and the unaltered phlogopite within a cleavage sheet. This mechanism produces the chlorite (clinochlore) ps. after phlogopite that preserve the external appearance of the original phlogopite crystal and rarely show any expansive separation of the sheets along cleavage planes. A second mechanism, by contrast, is incoherent and causes a significant increase in volume. At Pierrepont, these pseudomorphs generally have a silver-green bulging cleavage surface made up of many individual cleavage domains. The pseudohexagonal prism faces remain planar, but the basal pinacoid faces are curved outward. Both varieties of chlorite ps. after phlogopite occur at Pierrepont, but the coherent, unbulged variety is probably more common.

Figure 215. Clinochlore after phlogopite. 4.7 cm. Top of Hill. This pseudomorph formed by coherent alteration of phlogopite to clinochlore. Collected by Steve Chamberlain in 1975 and in his collection, #5045. *SCC*

Figure 216. Clinochlore after phlogopite. 7.7 cm. Middle Vein. This pseudomorph formed by incoherent or disruptive alteration of phlogopite to clinochlore. The center of the cleavage surface protrudes 3 mm above the rim. Collected by Steve Chamberlain in 2008 and in his collection, #19377. *SCC*

Pyrite ps. after Pyrrhotite. Although pyrite pseudomorphs after pyrrhotite are relatively common worldwide, they are unusual in the Precambrian rocks of northern New York because of the rarity of euhedral pyrrhotite crystals. Pyrrhotite crystals appear to be extremely rare at Pierrepont, having only been found one place on top of the hill where they are completely replaced by pyrite. Since these pseudomorphs were found embedded in calcite in close association with pyrite crystals, one assumes that the formation of the original pyrrhotite and the subsequent replacement by pyrite may have occurred during the original cooling of the pegmatite. As pyrrhotite cools, it releases sulfur and thereby drives the formation of pyrite. Further dissolution often results in direct replacement of the pyrrhotite by pyrite with a finely granular texture (Murowchick, 1992; Qian et al., 2011), which is what appears to have happened at Pierrepont.

Quartz ps. after Phlogopite. A new kind of pseudomorph was found in some abundance in a portion of the Phosphate Vein. Many of these pseudomorphs looked like pine cones covered in mud and only survived because they were altered to quartz. Quartz pseudomorphs after phlogopite have the same overall shape as the other phlogopite crystals from the vein, but some are expanded perpendicular to the cleavage plane. The two forms of pseudomorphs occurred side by side on the vein wall.

The pseudomorphs that are not expanded, but preserve the shape of the original phlogopite crystals with only minor dislocations along cleavage planes, are distinctly different in appearance from those that show exfoliation of mm-thick plates of phlogopite with empty inter-plate space, and may have a different mode of formation. One possibility is that all the quartz pseudomorphs after phlogopite had already altered to chlorite, either by the coherent mechanism described above, or by the incoherent, expansive mechanism, causing exfoliation. Quartz then would have replaced chlorite, rather than phlogopite directly. However, SEM examination of samples of

Figure 217. Pyrite after pyrrhotite, pyrite, quartz, talc. 1.5-cm crystal. Top of Hill. Collected by Mike Walter in 2003, now in the Chamberlain collection, #15567. MW

Figure 218. Pyrite after pyrrhotite, talc, quartz. 1.5-cm crystal. Top of Hill. Collected by Mike Walter in 2003, now in the Chamberlain collection, #15572. MW

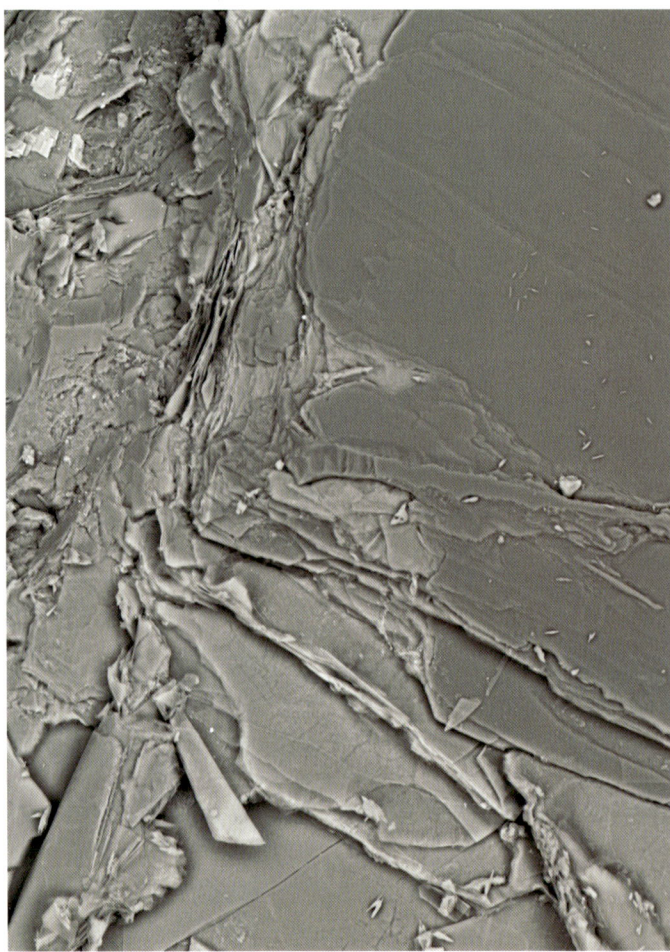

Figure 219. SEM image of partial replacement of phlogopite (darker gray) by quartz (lighter gray) along cleavage planes. 250-μm fov. SCC & DGB

Figure 220. Quartz after phlogopite. 2.8 cm. Phosphate Vein. Collected by Mike Walter in 2008, now in the Chamberlain collection, #19453. *SCC*

Figure 221. Quartz after phlogopite (right) and Chlorite after Phlogopite (left). 4.3 cm. Phosphate Vein. Collected by Mike Walter in 2008, now in the Chamberlain collection, #19176. *SCC*

phlogopite with some quartz replacement on the tips reveals that the quartz might directly replace phlogopite because coherent arrays of both phlogopite and quartz, but no chlorite, spanning cleavage sheets are observed. Of course, an alternative explanation is that partial replacement by chlorite was fully altered to quartz.

The exfoliated pseudomorphs would seem to require alteration from phlogopite to chlorite by the incoherent alteration mechanism to explain the exfoliation. For these unusual pseudomorphs, the order of events is likely to have been phlogopite → chlorite (with exfoliation) → quartz.

Consideration of the exfoliated pseudomorphs reveals hints about some of the geological history of the Phosphate Vein. The original phlogopite crystals were part of a complex, calcite-cored pegmatite that intruded the gneissic wall rock in Precambrian time, and essentially remained as such until the calcite core dissolved away. The dissolution of the core could have occurred slowly, even intermittently, over a long period of time beginning when the porous Potsdam Sandstone first eroded to the surface and continuing to the present. Clearly the exfoliation of the phlogopite crystals as they altered to chlorite must have occurred in the absence of the calcite core. Both the chloritization and further alteration to quartz might have occurred during the low-temperature, hydrothermal mineralization that also happened after the calcite core was gone. The spaces between exfoliated sheets contain small late-stage quartz, and some of the pseudomorphs have larger late-stage quartz crystals on their surfaces.

The seeming uniqueness of these pseudomorphs may reflect the particular complex geological history of the vein in which they were found. Clearly, interesting processes have been at work since the original emplacement of the pegmatite in the Precambrian.

Pseudomorphs after Diopside. There are four different kinds of pseudomorphs after diopside at Pierrepont: tremolite, quartz, talc, and uralite. All of these most likely formed during a period of retrograde metamorphism after the intrusion of the pegmatite.

Tremolite frequently overgrew diopside crystals epitactically at this occurrence. When the tremolite overgrowth is complete, the result is a tremolite ps. after diopside. The tremolite is typically dark green and is composed of many parallel fibers. If the pseudomorph is broken or if the encrustation is incomplete, it is possible to determine whether or not the diopside inside is unaltered.

Quartz relatively frequently replaces diopside crystals at various sites across the locality, especially the Top of the Hill, the Hillside, and the Phosphate Vein. Sometimes the quartz replaces only one end of a crystal and the other end is altered in some other manner. On Top of the Hill and Hillside sites, the original diopside was usually blocky in habit. In the Phosphate Vein, the original diopside crystals were prismatic with several different forms making up the termination. In all cases, the replacement of diopside by quartz ranges from an encrustation pseudomorph covering the surface to a complete internal replacement. The quartz is typically white to translucent gray and has an obvious fine-grained, pebbly replacement texture.

Talc occasionally coats or replaces diopside, forming talc ps. diopside, also known by the somewhat archaic varietal name rensselaerite. Usually these are encrustation pseudomorphs, but occasionally the original diopside is altered completely.

Uralite is a varietal name for an amphibole pseudomorph after pyroxene. At this locality, uralite is used to describe blocky diopside crystals that have been altered to a greater or lesser degree, resulting in a dull greenish-gray pseudomorph. XRD and SEM/EDS analyses showed that this uralite is often a mixture of talc, relict diopside, quartz, mica, and tremolite. However, individual specimens differ widely in the amounts of these constituents depending on the nature and degree of replacement.

Pseudomorphs after Scapolite. The scapolite at Pierrepont is intermediate in composition, but falls slightly in the marialite field of the marialite-meionite series. It is actually fairly rare, having been found only in a small area on top of the hill. Most of the scapolite has been altered to other minerals. In the two streamside veins, the Phosphate and Waddell Veins, sturdy, dark-green, sharply formed pseudomorphs made up of a mixture of clinochlore and muscovite are relatively common on the walls of the vein. On top of the hill, near, but not actually in, the Smoky Quartz Vein, pale greenish-gray pseudomorphs composed of a mixture of clinochlore and antigorite are relatively common. These are fragile and sometimes crumble when handled. Very rarely, dark green, translucent pseudomorphs of talc after scapolite are found at the same site. All three kinds of pseudomorphs after scapolite show clear tetragonal crystal forms.

The presence of unaltered marialite at some distance from the Smoky Quartz Vein on top of the hill and the consistent association of the pseudomorphs with veins that were mineralized by low-temperature hydrothermal solutions suggests that the pseudomorphs may have formed by reaction with those solutions, perhaps contemporaneously with the formation of the quartz pseudomorphs after phlogopite found in the Phosphate Vein.

Rare-Earth Minerals

All the rare-earth minerals tentatively identified from the Pierrepont locality have been found in minute crystals. Moreover, no rare-earth-elements (REE) have been detected, beyond parts-per-million concentrations, during analyses of the larger crystals of primary minerals that formed in the original pegmatite during the Precambrian. Our identifications are based on a combination of SEM/EDS analysis of chemistry and examination of external crystal

Figure 222. Uralite. 7.5 cm. Top of Hill. Collected by Mike Walter in 2013 and in his collection, #00700. *MW*

morphology. We have not confirmed the crystal structure with XRD for any of them largely because their microscopic size makes this very technically challenging.

The first REE mineral to be identified was allanite-(Ce), which occurs as mm-sized dark prisms on rhombohedral calcite crystals in open pockets found in lateral branches of the fracture zone that includes the Smoky Quartz Vein. Subsequently, synchysite-(Ce) was unexpectedly discovered in fractures in tourmaline crystals from the walls of the Phosphate Vein. Thereafter, an intentional search for REE minerals was conducted with samples from the Phosphate Vein and the Smoky Quartz Vein. An abundance of REE species was uncovered. It is likely that all the streamside veins have trace REE minerals, but this remains to be investigated.

The five REE minerals thus far identified and their sites of occurrence are shown in the table.

Species	Formula	Phosphate Vein	Smoky Quartz Vein
Allanite-(Ce)	$CaCeAl_2Fe_{2+}Si_3O_{12}(OH)$	x	x
Synchysite-(Ce)	$Ca(Ce,La,Nd)(CO_3)_2F$	x	x
Synchysite-(Y)	$Ca(Y,Ce,Nd,La)(CO_3)_2F$		x
Monazite-(Ce)	$(Ce,La,Nd,Th)(PO_4)$		x
Xenotime-(Y)	$Y(PO_4)$		x

Figure 223. Table of rare-earth minerals found. Only the Smoky Quartz Vein and the Phosphate Vein have been extensively surveyed.

The differences in occurrence may simply represent a sampling bias since the REE minerals in the Smoky Quartz Vein were all found together in massive chamosite and within calcite crystals. By contrast, those in the Phosphate Vein were scattered. Alternately, it could indicate that separate hydrothermal solutions with different REE compositions mineralized the two veins.

As discussed in more detail in the next section, the presence of REE minerals at several sites across this locality suggests a period of low-temperature hydrothermal mineralization was active (possibly one not previously recognized in the geological history of the region). At Pierrepont, deposition of the REE minerals was accompanied by barite, bornite, calcite, chalcocite, chalcopyrite, chamosite, microcline, pyrite, quartz, sphalerite, and talc crystals.

Periods of Mineralization: Precambrian and Late-Stage

As discussed in detail in chapter 2, the original minerals such as tourmaline, quartz, diopside, and phlogopite formed in the Precambrian as a pegmatite. Until now, the standard view of the remainder of the geologic history of the black tourmaline deposit consisted of a period of uplift and weathering followed by a period of subsidence and burial under a sequence of Paleozoic sedimentary rocks beginning with the Potsdam Sandstone. The unconformity between the Precambrian basement rocks and the Potsdam Sandstone represents about 500 million years of geological history. Finally, the region experienced a period of gradual uplift that continues into the present, during which most or all of the Paleozoic sediments weathered away. Much of the region around the tourmaline occurrence has Precambrian rocks at the surface, lightly covered here and there by patches of Potsdam Sandstone or glacial till from Pleistocene glaciation. Until recently, it was presumed that no additional mineralization other than simple chemical weathering occurred at Pierrepont in the past 500 million years.

Our detailed observations and analyses of the minerals present at the Pierrepont locality show that at least one period of additional mineralization has occurred. We believe these late-stage minerals were deposited by low-temperature, hydrothermal fluids that included both the REEs and other essential elements. Although the largest volume of late-stage mineralization consisted of quartz and calcite, five REE minerals, barite, microcline, chamosite, and talc crystals were also precipitated in open fractures. In the case of the Smoky Quartz Vein, the fractures were apparently open at the time of mineralization. In the case of the streamside veins, only the upper portions of the veins, where the calcite core had already dissolved away, were mineralized.

It is tempting to suggest that the sphalerite, pyrite, chalcocite, and bornite may have been deposited at the same time as the remobilization of the Precambrian sulfide ore and precipitation of similar minerals in fractures at the various zinc deposits of the nearby Balmat District. Similar sulfide mineralization in open fractures has been observed at the Selleck Road occurrences (Chamberlain et al., 2015) near West Pierrepont. Since some of these sulfides are within the late-stage calcite crystals in the Smoky Quartz Vein, the sulfide mineralization is likely of the same age.

REE minerals such as those found at Pierrepont have also been documented at the Rossie lead veins (Robinson et al., 2001)), the danburite locality south of Russell (Chamberlain et al, 2011), the Selleck Road occurrence (Chamberlain et al., 2015) near West Pierrepont, and the Route 30 road cut north of Long Lake (Richards & Robinson, 2000). At all these other occurrences, the REE minerals are also the result of late-stage, low-temperature mineralization and not derived locally from the deposit's primary minerals of Precambrian age. This is consistent with a regional period of REE-bearing low-temperature, hydrothermal mineralization.

Other than an $^{40}Ar/^{39}Ar$ date of 186 Ma obtained from adularia from the Rossie lead veins, we have no empirical way to establish the age of this REE mineralization. The Monteregian intrusions (117–138 mya) in nearby Quebec deposited numerous REE minerals at Mont Saint-Hilaire (Mandarino and Anderson, 1989), Oka (Treiman and Essene, 1985; Zurevinski and Mitchell, 2004), and other smaller nearby intrusives are possible sources of REE mineralization, as is a new, previously unidentified source.

LITERATURE CITED

Akavan, A. C. http://www.quartzpage.de/crs_habits.html.

Chamberlain, S. C., M. Lupulescu, D. G. Bailey, and D. M. Carlin Jr., "Classic danburite locality near Russell, St. Lawrence County, New York: New collecting and new research." *Rocks & Minerals* 86 (2011): 175–176.

Chamberlain, S. C., G. W. Robinson, M. R. Walter, and D. G. Bailey. "The Selleck Road tremolite and tourmaline locality, West Pierrepont, St. Lawrence County, New York. *Rocks & Minerals* 91 (2016): 116–130.

Clark, C. M., F. C. Hawthorne, and J. D. Grice. "Uvite, IMA 2000-030a." CNMNC Newsletter, April 2010, *Mineralogical Magazine* 74 (2010): 375–377.

Dana, E. S. *The System of Mineralogy of James Dwight Dana, 1837–1868.* New York: John Wiley & Sons, 1892.

Dietrich, R. V. *The Tourmaline Group.* New York: Van Nostrand Reinhold, 1985.

Dunn, P. J., D. Appleman, J. Nelen, and J. Norberg. "Uvite, a new (old) common member of the tourmaline group and its implications for collections." *Mineralogical Record* 8 (1977): 100–108.

Dyar, M. D., M. E. Taylor, T. M. Lutz, C. A. Francis, C. V. Guidotti, and M. Wise. "Inclusive chemical characterization of tourmaline: Mössbauer study of Fe valence and site occupancy." *American Mineralogist* 83 (1988): 848–864.

Grice, J. D. and T. S. Ercit. "Ordering of Fe and Mg in the tourmaline crystal structure: The correct formula." *Neues Jahrbuch für Mineralogie Abhandlungen* 165 (1993): 245–266.

Henry, D. J., M. Novak, F. C. Hawthorne, A. Ertl, B. L. Dutrow, P. Uher, and F. Pezotta. "Nomenclature of the tourmaline-supergroup minerals." *American Mineralogist* 96 (2011): 895–213.

Lupulescu, M. V. and M. Hawkins. "Tourmaline-supergroup minerals of New York—the revised nomenclature." Mineral News 29(1) (2013): 8–9, 15.

Mandarino, J. A. and V. Anderson. *Monteregian Treasures.* New York: Cambridge University Press, 1989.

Mullis, J., J. Dubessy, B. Poty, and J. O'Neil. "Fluid regimes during late stages of a continental collision: physical, chemical, and stable isotope measurements of fluid inclusions in fissure quartz from a geotraverse through the Central Alps, Switzerland." *Geochimica et Cosmochimica Acta* 58 (1994): 2239–2267.

Murowchick, J. B. "Marcasite inversion and the petrographic determination of pyrite ancestry." *Economic Geology* 87 (1992): 1141–1152.

Qian, G., F. Xia, J. Brugger, W. M. Skinner, J. Bei, G. Chen, and A. Pring. "Replacement of pyrrhotite by pyrite and marcasite under hydrothermal conditions up to 220°C: An experimental study of reaction textures and mechanisms." *American Mineralogist* 96 (2011): 1878–1893.

Rakovan, J. "Hemimorphism." *Rocks & Minerals* 82 (2007): 329–333.

Richards, R. P., and G. W. Robinson. "Mineralogy of the calcite-fluorite veins near Long Lake, New York." *Mineralogical Record* 31 (2000): 413–422.

Robinson, G. W., G. R. Dix, S. C. Chamberlain, and C. Hall. "Famous mineral localities: Rossie, New York." *Mineralogical Record* 32 (2001): 273-293.

Stalder, H. A. and A. Weisbrod. "Fluid inclusion studies in quartz from fissures of western and central Alps." *Schweizerische Mineralogische und Petrographische Mitteilungen* 54 (1974): 717–752.

Toksoy-Koksal, F., A. G. Turkmenoglu, and M. C. Goncuoglu. "Vermiculitization of phlogopite in metagabbro, Central Turkey." *Clays and Clay Minerals* 49 (2001): 81–91.

Treiman, A. H., and E. J. Essene. "The Oka carbonatite complex, Quebec: geology and evidence for silicate-carbonate liquid immiscibility." *American Mineralogist* 70 (1985):1101–1113.

Yan, Y-C., L. M. Anovitz, E. J. Essene, and D. R. Peacor. "Phlogopite-chlorite reaction mechanisms and physical conditions during retrograde reactions in the Marble Formation, Franklin, New Jersey." Contributions to *Mineralogy and Petrology* 88 (1984): 299–306.

Zurevinski, S. E., and R. H. Mitchell. "Extreme compositional variation of pyrochlore-group minerals at the Oka carbonatite complex, Quebec: Evidence of magma mixing?" *Canadian Mineralogist* 42 (2004): 1159–1168.

GLOSSARY

Adirondack blueline—A term denoting the boundary of the Adirondack Park.

anhedral—A mass that is one single crystallographic array of atoms inside, but which shows few if any poorly formed crystal faces.

anion—An atom or group of atoms with a negative charge.

apfu—atoms per formula unit.

backscattered electron imaging—A particular mode of operating a scanning electron microscope that is normally used to detect variations in chemical composition or the presence of multiple phases in the sample. Phases with an overall higher average atomic number will appear as brighter areas in the image.

botryoidal—A mineral aggregate with a bubbly surface, resembling a bunch of grapes.

calc-silicate rock—A regionally metamorphosed carbonate-rich (calcite and/or dolomite) rock containing significant calcium-rich silicate minerals (commonly diopside, wollastonite, anorthite, titanite, etc.).

coherent—Cohesiveness, or ability to hold together; in chemical alteration of a crystal to form a pseudomorph, a change in composition with only minor change in volume.

dipyramid—A crystal form consisting of two identical pyramids fastened at their common base, or terminating opposite ends of a prismatic crystal.

ditrigonal—In the hexagonal (trigonal) crystal system, general forms that do not align with the major crystallographic axes and have double the number of faces as those that do align, e.g. trigonal and ditrigonal pyramids, trigonal and ditrigonal prisms.

ductile deformation—The tendency for a rock to fold or behave plastically under heat and pressure rather than to fracture.

empirical formula—The simplest chemical formula for a substance; the one giving the simplest ratio of the various constituent atoms.

epitaxy/epitactic—The occurrence of one mineral species growing on the surface of another species in a special crystallographic orientation due to structural similarities in that plane or direction.

extensional faulting—Same as a normal fault, but which causes the crust to be extended.

fabric—The overall pattern, size and orientation of the individual mineral grains in a rock that, taken together, form its texture.

floater—A crystal that shows no point of attachment to a surface from which it grew.

foliation—The repeated, parallel wavy banding seen in metamorphic rocks caused by the preferred orientation of flat or platy mineral grains (e.g., micas) in response to directed pressure.

habit—The characteristic shape of a crystal that results from crystal growth, e.g., tabular, equant, prismatic.

hemimorphic—A term applied to crystals with no center of symmetry such as tourmaline, characterized by different crystal forms appearing on opposite terminations.

hexagonal—One of the six crystal systems having three equal-length axes 120 degrees apart and one longer or shorter six-fold axis perpendicular to the plane of the other three.

hydrothermal solution—A hot water solution containing dissolved minerals.

hydroxyl—The (OH)- anion.

intrusion—A younger igneous rock that cuts through (or "intrudes") a different, older rock.

LA-ICP-MS—An acronym for laser ablation, inductively coupled plasma, mass spectrometry; an analytical means by which trace element analyses are often performed.

lattice—The three-dimensional regular arrangement of atoms within a crystal.

metasedimentary rock—A metamorphic rock that formed by the recrystallization of a rock that was originally sedimentary.

monoclinic—One of the six crystal systems having three axes of different length, two of which are at 90 degrees and the third axis inclined to the plane of the other two.

morphology—The overall shape assumed by a crystal due to the relative development of the crystal forms present.

mylonite—A rock that has been so severely sheared and deformed by faulting and conditions of metamorphism that its original texture has been erased, and a new finely banded texture developed.

ocherous—Resembling ocher, or soft powdery brown or red iron oxides (goethite or hematite).

octahedral—Having the shape of an octahedron: eight identical equilateral triangles arranged as two square pyramids aligned base to base.

orogeny—A mountain-building event.

paleo-surface—An older, pre-existing surface.

Paleozoic—The time in Earth's history from approximately 540 to 250 million years ago.

pedion—A single crystal face in crystals without a center of symmetry.

pegmatite—An unusually coarse-grained intrusive igneous rock.

petrographic microscope—A microscope that uses polarized light to observe or measure optical properties of minerals, thereby identifying them.

pinacoid—A single pair of parallel, symmetrically equivalent faces.

pleochroism—Property of showing different colors in different crystallographic directions.

Precambrian—The time in Earth's history from approximately 4.6 billion to 540 million years ago.

prismatic—A crystal habit in which the crystal shows elongated growth parallel to a prism.

pseudomorph—The replacement of one mineral by another whereby the shape of the first mineral is retained.

pseudohexagonal—Having six sides that appear to form a regular hexagon, but actually don't. Most often four of the angles will be the same and two of the angles will be the same, but none will be 120 degrees.

pseudorhombohedral—Having three terminal faces on one end and three terminal faces on the other end of the crystal that are not actually at the correct angles for a rhombohedron, but look like one.

quartzofeldspathic—A rock consisting largely of quartz and feldspar(s).

radiaxial—Radiating outward from a common central point like spokes in a wheel.

rare-earth—Scandium, yttrium, and the lanthanides (those elements with atomic numbers 57 to 71) comprise the rare earth elements.

REE—An abbreviation for Rare-Earth Element(s).

retrograde metamorphism—That part of a metamorphic cycle characterized by decreasing heat and pressure, which follows prograde metamorphism.

rhombohedral—Having the shape of a rhombohedron: a crystal form common in the hexagonal (trigonal) crystal system with six rhombus-shaped, symmetrically equivalent faces (actually two sets of three), each set having three faces which are 120 degrees apart around the major crystal axis with a 60 degree rotation on the other end. Common in calcite-group minerals.

scalenohedral—A crystal form whose faces are all in the shape of scalene triangles.

SEM/EDS—An acronym for Scanning Electron Microscopy - Energy Dispersive Spectrometry. EDS refers to the way by which X-rays emitted by the elements in the sample are detected, enabling a rapid, non-destructive analysis for those elements with atomic number 6 or greater.

skeletal (crystal)—A crystal with faces that show negative or hopper-like growth.

stoichiometry—The quantitative proportions or ratios of elements in a compound.

subhedral—A rather poorly formed crystal, usually rounded or missing faces.

syncline—A down-warping or U-shaped fold in a rock.

tabular—Having a flat, planar, table-like crystal habit.

tectonic—A general term referring to the dynamics of the Earth's crust, such as mountain-building processes.

tetragonal—The crystal system with three mutually perpendicular axes, two of which are equal in length and the third longer or shorter than the other two.

trigonal—The division of the hexagonal crystal system with a three-fold principal axis.

uralitization—The replacement of a pyroxene (such as diopside) by an amphibole (such as tremolite).

vitreous—Having a glassy luster.

XRD—An acronym for X-ray Diffraction, a method of mineral identification based on a mineral's atomic structure.

ZAF correction—A computer program used to correct raw data obtained by electron microprobe analyses for atomic number effects, absorption, and fluorescence.